FATS IN FOOD TECHNOLOGY

FATS IN FOOD TECHNOLOGY

Dr. Sapna Kumari

Oxford Book Company
Jaipur ▪ New Delhi

Published By

Oxford Book Company

Regd. Off. "Bony Residency", Gate No. 2
Opp. Tilak Public School, Vishweswariya Nagar,
Gopalpura Road, Jaipur - 302018, India
Ph: 0141 - 2761280, Telefax: 2761381
email - oxfordbookjpr@gmail.com
website - oxfordbookcompany.com

Delhi Office: 102, 1st Floor, Satyam House,
4327/3, Ansari Road, Darya Ganj,
New Delhi - 110002, India
Ph: 011-45652440

Distributed By

KEYA INTERNATIONAL PRESS LLP
E-mail: keyainternationalpressllp@gmail.com

Published in India, 2025

ISBN : 9789350303757

Typeset by Shivangi Computers, Jaipur

Printed and bound in India by Thomson Press Ltd.,
New Delhi

Contents

Preface

Fat is a type of nutrient. You need some fat in your diet but not too much. Fats give you energy and help your body absorb vitamins. Dietary fat also plays a major role in your cholesterol levels. Fats are made when each of these three -OH groups reacts with a fatty acid. The resulting fats are called triglycerides. Because of their preponderant aliphatic structure, fats are hydrophobic, generally soluble in organic solvents but generally insoluble in water. Fats made up of shorter chain fatty acids are usually liquid at room temperature, whereas the longer chain fats will be solid. Some ambiguity in terminology arises because the words "oil", "fat", and "lipid" are often used interchangeably. Of these lipid is the general term, because a lipid is not necessarily a triglyceride. Oil is the term usually used to refer to fats that are liquids at normal room temperature, while fat is usually used to refer to fats that are solids at normal room temperature. Fat is important foodstuff for many forms of life, and fats serve both structural and metabolic functions. They are necessary part of the diet of most heterotrophs (including humans). Some fatty acids that are set free by the digestion of fats are called essential because they cannot be synthesized in the body from simpler constituents. There are two essential fatty acids (EFAs) in human nutrition: alpha-linolenic acid (an omega-3 fatty acid) and linoleic acid (an omega-6 fatty acid). Other lipids needed by the body can be synthesized from these and other fats. Fats and other lipids are broken down in the body by enzymes called lipases produced in the pancreas.

Annual production and consumption of oils and fats is about 119 million tonnes and rising steadily at a rate of 2–6 million tonnes per year. This is required to meet the demand, which also grows at around this rate, partly as a consequence of increasing population but more because of increasing income, especially in developing countries. Around 14% of current oil and fat production is used as starting material for the oleochemical industry and around 6% is used as animal feed (and indirectly therefore as human food). The remaining 80% is used for human food—as spreads, frying oil, salad oils, cooking fat, etc. Fats and oils provide important functions in the food and beverage products, from texture and shelf-life to flavour. Food product developers expect edible oils to perform in many different applications and consumers want them to deliver healthy benefits. New technologies are creating new lipids to meet a wide range of requirements. From the latest research on health effects to improved functionality to the hottest trends, this fats and oils category covers everything relevant in the food industry. Animal-based fats were once the only trans fats consumed, but by far the largest amount of trans fat consumed today is created by the processed food industry as a side-effect of partially hydrogenating unsaturated plant fats (generally vegetable oils). These partially-hydrogenated fats have displaced natural solid fats and liquid oils in many areas, notably in the fast food, snack food, fried food and baked goods industries. Partially hydrogenated oils have been used in food for many reasons. Partial hydrogenation increases product shelf life and decreases refrigeration requirements. Many baked foods require semi-solid fats to suspend solids at room temperature; partially hydrogenated oils have the right consistency to replace animal fats such as butter and lard at lower cost. They are also an inexpensive alternative to other semi-solid oils such as palm oil.

This basic reference book provides comprehensive and detailed coverage of food oils and fats in a practical, how-to format. The book integrates all aspects of food oils and fats from chemistry to food processing to nutrition.

—Dr. Sapna Kumari

Chapter 1

Physical Properties of Fats in Food

Introduction

Fat is one of the three main macronutrients: fat, carbohydrate, and protein. Fats are a wide group of compounds whose basis is in long-chain organic acids, called fatty acids. More particularly fats are esters of such organic acids formed with the alcohol glycerol. Glycerol is a triol, meaning that it has three chemically active -OH (hydroxyl) groups. Fats are made when each of these three -OH groups reacts with a fatty acid. The resulting fats are called triglycerides. Because of their preponderant aliphatic structure, fats are hydrophobic, generally soluble in organic solvents but generally insoluble in water. Fats made up of shorter chain fatty acids are usually liquid at room temperature, whereas the longer chain fats will be solid. Some ambiguity in terminology arises because the words "oil", "fat", and "lipid" are often used interchangeably. Of these lipid is the general term, because a lipid is not necessarily a triglyceride. Oil is the term usually used to refer to fats that are liquids at normal room temperature, while fat is usually used to refer to fats that are solids at normal room temperature.

Fat is important foodstuff for many forms of life, and fats serve both structural and metabolic functions. They are necessary part of the diet of most heterotrophs (including humans). Some fatty acids that are set free by the digestion of fats are called essential because they cannot be synthesized in the body from simpler constituents. There are two essential fatty acids (EFAs) in human nutrition: alpha-linolenic acid (an omega-3 fatty acid) and linoleic acid (an omega-6 fatty acid). Other lipids needed by the body can be synthesized from these and other fats. Fats and other lipids are broken down in the body by enzymes called lipases produced in the pancreas.

Fats and oils are categorized according to the number and bonding of the carbon atoms in the aliphatic chain. Fats that are saturated fats have no double bonds between the carbons in the chain. Unsaturated fats have one or more double bonded carbons in the chain. The nomenclature is based on the non-acid (non-carbonyl) end of the chain. This end is called the omega end or the n-end. Thus alpha-linolenic acid is called an omega-3 fatty acid because the 3rd carbon from that end is the first double bonded carbon in the chain counting from that end. Some oils and fats have multiple double bonds and are therefore called polyunsaturated fats. Unsaturated fats can be further divided into cis fats, which are the most common in nature, and trans fats, which are rare in nature. Unsaturated fats can be altered by reaction with hydrogen effected by a catalyst. This action, called hydrogenation, tends to break all the double bonds and makes a fully saturated fat. To make vegetable shortening, then, liquid *cis*-unsaturated fats such as vegetable oils are hydrogenated to produce saturated fats, which have more desirable physical properties e.g., they melt at a desirable temperature (30–40 °C), and store well, whereas polyunsaturated oils go rancid when they react with oxygen in the air. However, trans fats are generated during hydrogenation as contaminants created by an unwanted side reaction on the catalyst during partial hydrogenation. Consumption of such trans fats has shown to increase the risk of coronary heart disease

Saturated fats can stack themselves in a closely packed arrangement, so they can solidify easily and are typically solid at room temperature. For example, animal fats tallow and lard are high in saturated fatty acid content and are solids. Olive and linseed oils on the other hand are unsaturated and liquid.

Fats serve both as energy sources for the body, and as stores for energy in excess of what the body needs immediately. Each gram of fat when burned or metabolized releases about 9 food calories (37 kJ = 8.8 kcal). Fats are broken down in the healthy body to release their constituents, glycerol and fatty acids. Glycerol itself can be converted to glucose by the liver and so become a source of energy.

***Trans* (Elaidic acid)**	***Cis* (Oleic acid)**	***Saturated* (Stearic acid)**
Elaidic acid is the principal *trans* unsaturated fatty acid often found in partially hydrogenated vegetable oils.[36]	Oleic acid is a *cis* unsaturated fatty acid making up 55–80% of olive oil.[37]	Stearic acid is a saturated fatty acid found in animal fats and is the intended product in full hydrogenation. Stearic acid is neither *cis* nor *trans* because it has no carbon-carbon double bonds.

There are many different kinds of fats, but each is a variation on the same chemical structure. All fats are derivatives of fatty acids and glycerol. The fat molecules are called triglycerides (triesters of glycerol). Three chains of fatty acid are bonded to each of the three -OH groups of the glycerol by the reaction of the carboxyl end of the fatty acid (-COOH) with the alcohol. HOH (water) is eliminated and the carbons are linked by an -O- bond through dehydration synthesis. This process is called esterification and fats are therefore ester | esters. As a simple visual illustration, if the kinks and angles of these chains were straightened out, the molecule would have the shape of a capital letter E. The fatty acids would each be a horizontal line; the glycerol "backbone" would be the vertical line that joins the horizontal lines. Fats therefore have "ester" bonds.

The properties of any specific fat molecule depend on the particular fatty acids that constitute it. Fatty acids form a family of compounds are composed of increasing numbers of carbon atoms linked into a zig-zag chain (hydrogen atoms to the side). The more carbon atoms there are in any fatty acid, the longer its chain will be. Long chains are more susceptible to intermolecular forces of attraction (in this case, van der Waals forces), and so the longer ones melt at a higher temperature (melting point).

Fatty acid chains differ by length, often categorized as short to very long.

- Short-chain fatty acids (SCFA) are fatty acids with aliphatic tails of fewer than six carbons (i.e. butyric acid).
- Medium-chain fatty acids (MCFA) are fatty acids with aliphatic tails of 6–12 carbons, which can form medium-chain triglycerides.
- Long-chain fatty acids (LCFA) are fatty acids with aliphatic tails 13 to 21 carbons.
- Very long chain fatty acids (VLCFA) are fatty acids with aliphatic tails longer than 22 carbons

Any of these aliphatic fatty acid chains may be glycerated and the resultant fats may have tails of different lengths from very short triformin to very long, e.g., cerotic acid, or *hexacosanoic acid*, a 26-carbon long-chain saturated fatty acid. Long chain fats are exemplified by tallow (lard) whose chains are 17 carbons long. Most fats found in food, whether vegetable or animal, are made up of medium to long-chain fatty acids, usually of equal or nearly equal length. Many cell types can use either glucose or fatty acids for this energy. In particular, heart and skeletal muscle prefer fatty acids. Despite long-standing assertions to the contrary, fatty acids can also be used as a source of fuel for brain cells.

Importance for Living Organisms

Fats are also sources of essential fatty acids, an important dietary requirements. They provide energy as noted above. Vitamins A, D, E, and K are fat-soluble, meaning they can only be digested, absorbed, and transported in conjunction with fats.

Fats play a vital role in maintaining healthy skin and hair, insulating body organs against shock, maintaining body temperature, and promoting healthy cell function.

Fat also serves as a useful buffer towards a host of diseases. When a particular substance, whether chemical or biotic, reaches unsafe levels in the bloodstream, the body can effectively dilute—or at least maintain equilibrium of—the offending substances by storing it in new fat tissue. This helps to protect vital organs, until such time as the offending substances can be metabolized and/or removed from the body by such means as excretion, urination, accidental or intentional bloodletting, sebum excretion, and hair growth.

Adipose Tissue

In animals, adipose, or fatty tissue is the body's means of storing metabolic energy over extended periods of time. Adipocytes (fat cells) store fat derived from the diet and from liver metabolism. Under energy stress these cells may degrade their stored fat to supply fatty acids and also glycerol to the circulation. These metabolic activities are regulated by several hormones (e.g., insulin, glucagon and epinephrine).

The location of the tissue determines its metabolic profile: visceral fat is located within the abdominal wall (i.e., beneath the wall of abdominal muscle) whereas "subcutaneous fat" is located beneath the skin (and includes fat that is located in the abdominal area beneath the skin but above the abdominal muscle wall). Visceral fat was recently discovered to be a significant producer of signaling chemicals (i.e., hormones), among which several are involved in inflammatory tissue responses. One of these is resistin which has been linked to obesity, insulin resistance, and Type 2 diabetes. This latter result is currently controversial, and there have been reputable studies supporting all sides on the issue.

Basic Physical Properties of Fat Crystals

Fats are the main structural components in many food products such as chocolate, confectionery coatings, dairy products, butter, cream shortenings, margarine, and spreads. The sensory characteristics of fat-structured materials such as spreadability, hardness, and mouth feel are highly dependent on the structure of the underlying fat crystal network. This fat crystal network is built by the interaction of polycrystalline fat particles. The amount, geometry, and spatial distribution of solid fat crystals as well as their interactions at different levels within the network all affect the rheological properties of fats and fat-structured food products. Fat crystallization largely determines consistency, physical stability, visual appearance, and eating properties.

Crystals

A crystal consists of a material in a solid state in which the building entities—molecules, atoms or ions—are closely packed so that the free energy of the material is at minimum. Crystals show enormous variation in external shape or habit caused by variation in the growth rate of the various faces of a crystal, which rates often depend on the composition of the solution. Corners and edges are rounded; curved faces can appear in large crystals; some needlelike crystals have a slight twist; some faces can grow faster than others. A noncrystalline solid is often referred to as an amorphous solid. Whether a material is crystalline or not can be established by x-ray diffraction. X-rays have a very small wavelength, of the order of 0.1 nm, which implies that individual atoms may cause scattering. If the atoms (or small molecules) occur at regular distances, sharp diffraction maxima occur, and the crystal structure can be derived from the diffraction pattern.

Crystallization

Many of the sensory attributes such as spreadability, mouthfeel, snap of chocolate, texture, etc., are dependent on the mechanical strength of the underlying fat crystal network. In addition to this obvious industrial importance, fat crystal networks form a particular class of soft materials, which demonstrate a yield stress and viscoelastic properties, rendering these plastic materials. The levels of structure in a typical fat network are defined as the fat crystallizes from the melt. The growth of a fat crystal network can be visualized thus: the triacylglycerols present in the sample crystallize from the melt into particular polymorphic/polytypic states. These crystals grow into larger microstructural elements (H–6 mm) which then aggregate via a mass- and heat-transfer limited process into larger microstructures (H–100 mm). The aggregation process continues until a continuous three-dimensional network is formed by the collection of microstructures. Trapped within this solid network structure is the liquid phase of the fat.

The crystallization process consists of two steps: nucleation and crystal growth. Nucleation can be described as a process in which molecules come into contact, orient and interact to form highly ordered structures, called nuclei. Crystal growth is the enlargement of these nuclei. Nucleation and crystal growth are not mutually exclusive: nucleation may take place while crystals grow on existing nuclei.

Nucleation can only be achieved via supersaturation or supercooling. A solution is supersaturated if it contains more of a component than can be theoretically dissolved within it at a particular temperature. Supercooling refers to the degree to which the solution is cooled with respect to the melting temperature of the crystallized solution. It is very difficult to determine the parametres of supersaturation and supercooling for a crystallizing system, and therefore, as a good approximation, in practice only supercooling is usually considered for crystallization of triacylglycerol molecules from the melt.

When the temperature of a fat melt is decreased below its maximum melting temperature, it becomes supersaturated in the higher-melting triacylglycerol species present in the mixture.

This so-called undercooling or supercooling represents the thermodynamic driving force for the change in state from liquid to solid. Fats usually have to be undercooled by at least

5-10°C before they begin to crystallize. For a few degrees below the melting point, the melt exists in a so-called metastable region. In this region, molecules begin to aggregate into tiny clusters called embryos. At these low degrees of undercooling, embryos continuously form and breakdown, but do not persist to form stable nuclei. The energy of interaction between triacylglycerol molecules has to be greater than the kinetic energy of the molecules in the melt so as to overcome Brownian effects. For these flexible molecules, it is not sufficient to simply interact; molecules have to adopt a specific conformation in order to form a stable nucleus. The adoption of this more stable conformation is relatively slow, thus explaining the existence of a metastable region. As the undercooling is increased (i.e., at lower temperatures) stable nuclei of a specific critical size are formed.

When fats are cooled down from a high temperature (so completely melted fat) to temperatures below the melting point, crystals are formed; this can occur in different crystal modifications:

- Alpha-solids, formed when fat is quickly cooled down (seconds); these solids are unstable and will convert in more stable form as

- Betha-prime solids.
- Betha solids; these are the most stable modification.

Cooling of chocolate (with cocoa butter or cocoa butter equivalents) is more delicate for getting the right crystal form; tempering is then applied to create the right crystal form and to speed up the crystallisation.

In the margarine production undercooling is applied by rapid cooling in SSHE (surface scraped heat exchanger); then residence time has to be given to convert alpha solids in betha prime solids and to make more betha-prime solids.

Recrystallisation can occur; an example is crystallised partially hardened Rapeseed or Canola. These fats, although present in the betha-prime solids form, can recrystallise (esp. at variation of temperature) into betha form (resulting in large crystals in needle form). This recrystallisation process can be diminished or retarded by making the fat blend more complex e.g. with higher levels of C16.0 in a mainly C18-composition.

The crystallisation rate of a fat depends on:

- Rate of cooling.

In drum cooler or SSHE the cooling rate is fast; in a few seconds the melted fat can be cooled far below the melting point.

If cooling in a tank is applied (sometimes with fractionation) then the cooling time can take hours.

- The degree of undercooling, also depending on the rate of cooling.
- The type of triglycerides.

The alpha solids are formed rapidly in all cases that fast cooling is applied but the conversion into betha prime solids strongly depends on the type of triglycerides present.

- The steepness of the N-line or actually the amount of triglycerides crystallised at a temperature change.

Polymorphic Structures of Fats

Food fat provides taste, consistency, and helps us feel full. Fat is a major source of energy for the body, and aids in the absorption of lipid soluble substances including vitamins A, D, E, and K. Dietary fat is essential for normal growth, development, and maintenance, and serves a number of important functions. Increasing evidence indicates that fatty acids and their derived substances may mediate critical cellular events, including activation and expression of genes, and regulation of cellular signaling.

When and how humans learned to use fats and oils is unknown, but it is known that primitive people in all climates used them for food, medicine, cosmetics, lighting, preservatives, lubricants, and other purposes. The use of fats as food was probably instinctive, whereas the other applications most likely resulted from observations of their properties and behaviour under various environmental conditions. More than likely, the first fats used by humans were of animal origin and were separated from the tissue by heating or boiling. Recovery of oil from small seeds or nuts required the development of more advanced methods of processing, such as cooking, grinding, and pressing processes.

The total global oil and fat market is a huge economic factor. The rise of affluence in developing countries, this market is increasing and can be expected to increase further. The main fats commonly consumed are vegetable oils and fats, dairy fat and fats derived from animals, e.g. lard, tallow and fish oil.

Refining edible oils such as neutralization, bleaching, and deodorization, has been practiced for just over a century, but it has had a great impact on eating habits. Whereas the refining processes have increased the availability of sufficiently palatable oils, the oil modification.

Polymorphism

An important way to characterize fats and oils is through the predominant crystalline phase, or polymorph, that tend to form upon crystallization. When the same ensemble of molecules can pack in different arrangements on crystallization, depending on the processing conditions, the substance is said to demonstrate polymorphism. The different polymorphic states of a particular substance often demonstrate quite different physical properties (such as melting behaviour and hardness), but on melting yield identical liquids.

Polymorphism is the ability of long-chain compounds such as fatty acids to exist in more than one crystal form, and this results from different patterns of molecular packing in the crystals. Triacylglycerols may occur in three main forms, namely, α, β', and β in order of increasing stability and melting point. When fats are cooled, crystals of a lower melting form may be produced. These may change slowly or rapidly into a more stable form. The change is monotropic, that is, it always proceeds from lower to higher stability.

Polymorphism results in the phenomenon of multiple melting points. When a fat is crystallized in an unstable form and heated to a temperature slightly above its melting point, it may resolidify into a more stable form. The polymorphs differ in stability, melting point, melting enthalpy, and density. The α-polymorph is the least stable and has the lowest melting point, melting enthalpy, and density. The β-polymorph is the most stable and has the highest melting point, melting enthalpy, and density. The β2 -polymorph has intermediate properties.

Under rapid cooling conditions, triacylglycerol molecules usually crystallize in metastable polymorphic forms, which subsequently transform into polymorphs of higher stability. On the other hand, at slow cooling rates, triacylglycerol molecules of similar chain lengths have time to associate with each other in more stable geometrical arrangements, resulting in the formation of a more stable polymorphic form. Due to the dependence of fat crystallization on the degree of undercooling and the cooling rate used, different results will be observed when using different cooling rates.

Tempering

Before its solid fat content can be determined, the fat must be exposed to a prescribed temperature profile: first it has to be melted completely to destroy all traces of crystals, and then cooled to achieve virtually complete crystallization, and finally it has to be held at the measuring temperature to come to equilibrium at that temperature. Sometimes, depending on the fat used, an extra step is introduced where the fat is held at a particular temperature, which

is not the measuring temperature. This step is referred to as a tempering step. For confectionery fats, a tempering step of 40 hours at 26°C is mentioned in the standard methods to ensure that cocoa butter and similar fats like cocoa butter equivalents (CBEs) are converted to their β-polymorph before the SFC is measured.

Tempering is a technique of controlled pre-crystallization employed to induce the most stable solid form of cocoa butter, a polymorphic fat in finished chocolates. The process consists of shearing chocolate mass at controlled temperatures to promote crystallization of triacylglycerols in cocoa butter to effect good setting characteristics, foam stability, demoulding properties, product snap, contraction, gloss and shelf-life characteristics. Time–temperature protocols and shearing are employed to induce nucleation of stable polymorphs with the formation of three-dimensional crystal network structure influencing the microstructure, mechanical properties and appearance of products. The crystal network organization and the polymorphic state of the triacylglycerols crystals as affected by the crystallization conditions are major factors determining rheological and textural properties of crystallized triacylglycerols systems.

Solid Fat Content

The solid fat content (SFC) is a measure of the percentage of solid, crystalline fat in a sample at a selected temperature. Often, the SFC is measured at selected points within a temperature range. A measure of the SFC can be determined by a variety of methods: dilatometry, pulsed nuclear magnetic resonance (p-NMR), or differential scanning calorimetry (DSC). The method used and differences in the way it is executed can seriously affect the final result.

Fat Design

Each application area requires its proper fat. The specifications of the fat depend on: recipe, equipment, procedure, temperature of fat and other ingredients, ambient temperature, storage and distribution temperature of the final product. Some conditions to attend a satisfactory fat design must be the compatibility among the components of the mixture: equivalent thermal properties (solid fat content, melting point and range); similar molecular size, shape and packing (to allow isomorphous replacement or formation of a single lattice unit in mixtures); similar polymorphism (transformation from stable to unstable forms should occur as readily for binary mixtures as with individual components).

Processing

Edible fats and oils have been separated from animal tissues, oilseeds, and oil-bearing fruits for thousands of years. The combined largest source of vegetable oils is the seeds of annual plants grown in relatively temperate climates. The oilseeds are processed by expeller or screw press extraction, by prepress solvent extraction, or bay expander–solvent extraction.

A second source of vegetable oil is the oil-bearing tree fruits and kernels. Oil-bearing fruits are pressed to obtain oil, sometimes after drying or sterilizing, or are cold pressed to preserve flavour and odour. Animal tissues may be wet- or dry-rendered (cooking processes) to separate the fats. Edible meat fats are supplied by lard from pigs, tallow from cattle and sheep, and milk fat or butter from cows. After recovering, fats and oils can be physically and/or chemically refined.

Chemical refining removes most impurities with an alkaline solution, whereas physical refining removes them by distillation.

Cocoa Butter Crystallisation

Cocoa butter plays a significant role in the overall performance of chocolate. Fat is the continuous phase of chocolate and it is the cocoa butter that provides the snap, gloss and pleasant, cooling mouthfeel.

To obtain these desirable qualities from the cocoa butter, proper crystallisation is crucial. Incorrectly produced chocolate will have the wrong melting profile, warmer mouthfeel, poor gloss and a very short shelf life. It also will be subject to bloom formation. The difficulty in obtaining the correct crystallisation lies in the polymorphism of the fat. The composition of the cocoa butter -depending on origin, harvest and processing- as well as the chocolate recipe, will affect the polymorphic characteristics of the product.

Cocoa Butter Polymorphism

Despite numerous structures presented for the principal cocoa butter TAGs, StOSt, POSt and POP, it must be stressed that no structure has yet been determined. To date, the structures presented for these TAGs are suggestions only. Many clues have been gathered from x-ray diffraction and spectroscopic methods but work is hampered by the difficulty in obtaining a single crystal.

These three main TAGs themselves have a complicated polymorphism so it is no surprise that cocoa butter's polymorphism is also complicated. However, the fact that cocoa butter is a mixture of TAGs adds another layer of complexity.

Wille & Lutton are often credited with the discovery of the complete polymorphic scheme, naming the polymorphs using Roman numerals (I-VI) in order of increasing melting point.

The b phases have melting points anywhere between 30 and 36°C, depending on composition. The b_{VI} melting point usually being about 2°C higher than that for b_{V}. Note that the b_{V} melting point of a hard cocoa butter can be higher than the b_{VI} melting point of a softer cocoa butter. Similarly, it is possible for the b' melting point of a hard cocoa butter to be higher than that of the b form of a soft cocoa butter. As an identification of the polymorphic form, therefore, melting points are not reliable. Only x-ray diffraction provides an unequivocal means to determine the polymorphic form.

Crystallisation of Cocoa Butter

The different polymorphs not only have different melting points, but will crystallise at different temperatures. Moreover, different components in the cocoa butter may crystallise independently, depending on their solubility in the remaining liquid phase.

Crystallisation begins with nucleation, of which two types of nucleation can be distinguished: heterogeneous and homogeneous. The former means that crystal growth is initiated on "foreign" particles. These may be the sugar, cocoa powder particles, emulsifiers or even dust. Crystals from a different polymorph might be considered heterogeneous nuclei.

Homogeneous nucleation occurs without such "foreign" particles and requires significant super-cooling (cooling below the melting point). It is important to note that Sat-O-Sat TAGs will grow easily on crystals with similar TAG structure and much less on other TAGs, such as trisaturated ones, whatever the polymorph. Increasing the amount of fully saturated TAG in cocoa butter does not increase overall crystallisation performance, but increase of StOSt (or even BOB) does.

Although the crystallisation of cocoa butter depends on the formation of nuclei, temperature and physical motion (e.g. stirring) have an influence too.

There are two counteracting phenomena. Higher melting, more stable polymorphs have a higher activation energy for crystallisation and so require more time for crystallisation. On the other hand, they have a greater degree of super-cooling at any given temperature, compared to less stable polymorphs.

Effectively, it is the temperature that decides which polymorph will form. It is only when cooling is so slow that higher polymorphs find time to crystallise that lower polymorphs are prevented from crystallising.

Under static conditions, only a crystallises quickly (within minutes) straight from the melt, at temperatures below 20° C. In fact, g (sub-a) will also crystallise quickly if a temperature below 0°C is achieved following rapid (>2°C/min) cooling.

At any cooling rate the crystallisation of g will be accompanied by partial crystallisation of, and/or transformation to, a. At temperatures between 21 and 28°C, b' will slowly crystallise.

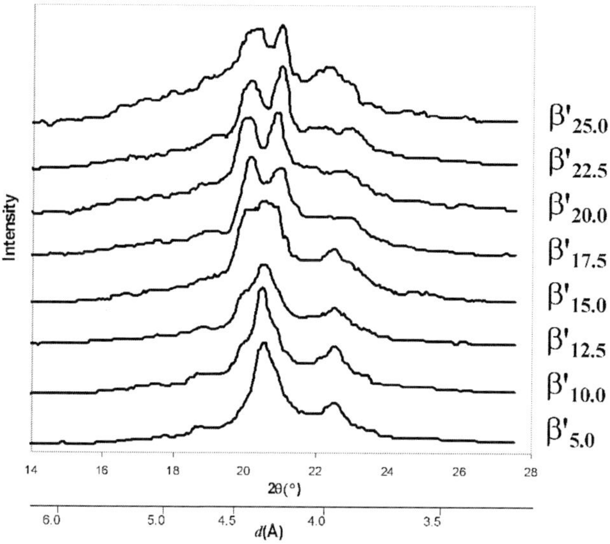

At higher temperatures the cocoa butter will remain liquid. Experiments maintaining cocoa butter at 30°C for several months yielded a minor part of the highest melting TAG floating as solids in an otherwise completely liquid fat.

Phase Transformations

It is a characteristic of fat polymorphism that polymorphic transformation is monotropic, taking place in one direction, from less stable to more stable. Once a fat is in a more stable phase, it must be dissolved or melted in order to regain a less stable phase. In general, polymorphs are easier formed by transformation from a lower polymorph than by crystallisation straight from the melt.

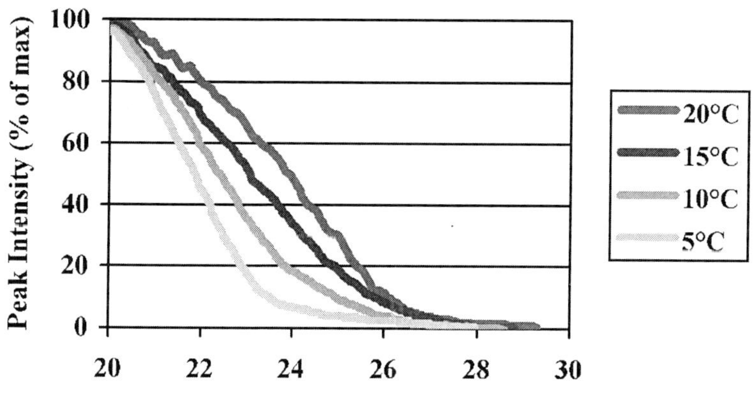

Depending on the temperature, polymorphs will generally transform to the more stable forms after some time. The a phase will transform into a b' phase, which will, in turn, transform into b. Depending on the temperature at which the a transforms into b', or the b' crystallises by itself, the physical properties (melting point, diffraction pattern) will differ.

For example, with no difficulty whatsoever, it is possible to generate 7 or more different b' polymorphs, each with its own melting profile and x-ray diffraction pattern. A thorough examination of this phase range demonstrates it cannot be reduced to simple combinations of two different phases existing in various ratios; it is truly a continuum of b' polymorphs.

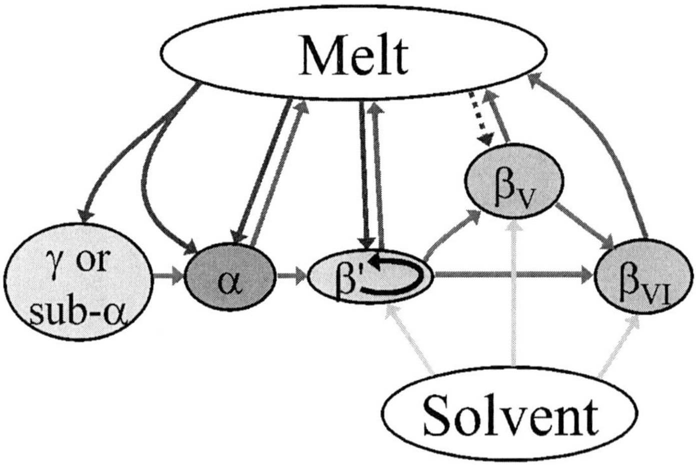

Depending on temperature, the b' will eventually transform into b. At lower temperatures (up to about 22°C), b' will transform into b_V. At higher temperatures (from about 26°C), b' will transform directly into the b_{VI} form. At intermediate temperatures, a mixture of b_V and b_{VI} will be formed.

Without seeding of some description, it is impossible to crystallise the b polymorphs directly from the melt. Suitable seeds are fat crystals, preferably in the desired polymorphic state and built from TAGs which are similar to those in cocoa butter. Thus b-cocoa butter seeds or b-StOSt seeds will accelerate crystallisation much more than b'-cocoa butter seeds. These latter, however, still will be much faster than any seed of tripalmitin or tristearin, even though these have much higher melting points.

A special type of seeding, or rather of having the right nuclei for fast crystallisation, is the memory effect. If cocoa butter in the b phase is melted (no visible solid), but remains below 40°C, it maintains a "memory" of the crystal structure. On subsequent cooling, the crystallisation speed of the b phase is increased, apparently forming from the melt.

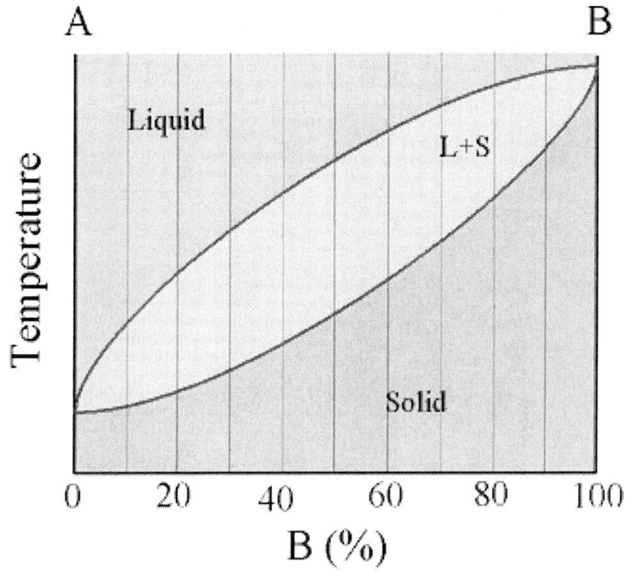

This temperature of 40°C applies irrespective of cocoa butter composition. That is, whether the cocoa butter has a melting point of 32°C or 36°C, the memory effect for both is lost at 40°C but not below. This implies that the invisible nuclei have similar compositions, irrespective of the cocoa butter source.

Concept of Crystallites as Individuals

A key to understanding phase transformations, the existence of a b' phase-range, memory effects and other "mysteries" of cocoa butter is the phenomenon of mixed crystals.

If we consider a simple phase diagram in which there is complete miscibility in both liquid and solid phases, we can follow what happens on cooling of a mixture. If we cool a molten mixture with 30% B from (1) we eventually reach the liquidus at (2). At this point crystallisation begins. However, the composition of the crystallising material is read off by taking a horizontal (isothermal) line across to (3) and finding the composition by dropping a vertical line to (4), i.e. about 73% B. During subsequent cooling, the composition of the crystallised material will follow the solidus line down to (5), while the composition of the remaining liquid will follow the liquidus down to (6).

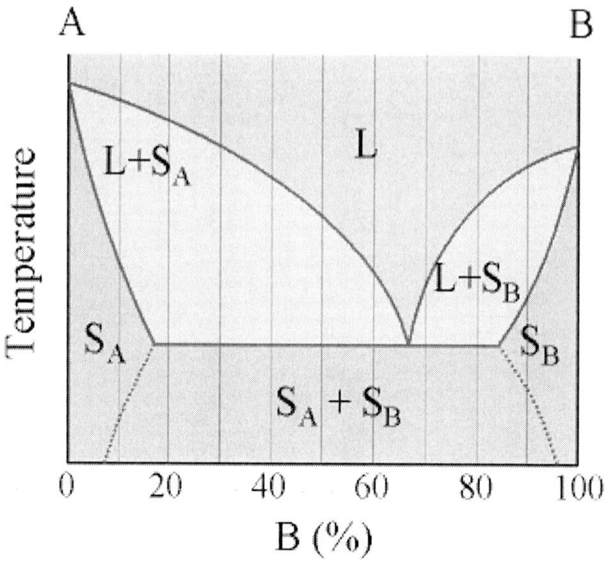

The process described above assumes that the system remains at equilibrium at all times and that there is sufficient time for the solid phase to remain homogenous. In practice, of course, this never happens. Frequently crystallisation is rapid meaning that there is no time for the solid phase to equilibrate to the thermodynamically stable composition.

Crystallisation can be imagined to take place in several stages, building up each crystal like the skins of an onion. In reality, the composition probably will vary continuously from the centre to the outside.

Of course, the crystals or crystallites that form will be of varying sizes as well as varying compositions.

The phase behaviour of binary combinations of fats need not be as simple. If partial immiscibility occurs, as it commonly does with TAGs. This leads to much more complex crystallisation.

The TAGs present in cocoa butter have a range of melting points. These relate to differences in saturation and chain length. When liquid cocoa butter is cooled, each of these TAG will perceive a different kind of super-cooling. It seems obvious that the "higher melting" TAG will crystallise first, but remember that we must think in terms of the phase diagram. The first solid will, of course, be rich in the higher melting TAGs but will also include some of the softer TAGs. A group of molecules will stick together to form an embryo (a nucleus too small to survive). Most embryos will dissolve quickly again, but some will grow. Around the kernel of this embryo, different TAG may join. In such a process, a nucleus and later a crystallite will grow with an individual composition. Numerous crystallites will develop, differing slightly in composition due to micro-fluctuations of composition and temperature in the liquid and due to formation at slightly different times. Eventually, each crystallite will have a unique composition, melting point and structure related to that particular composition. The overall melting range and polymorphic form of the solidified cocoa butter will be the average of all those individual crystallites.

Since the force driving crystallisation is affected by changes in temperature, variation in crystallite composition will occur. This could be the cause of the phenomenon of the b' phase range variation in melting point and diffraction pattern noted above.

Chapter 2

Bakery Fats

Never before has there been such a choice of fats in the chiller compartment of the supermarket. A recent survey found 20 different varieties ranging from traditional butter with 81.5% fat to a low fat spread containing only 40% fat. Between these there are polyunsaturated spreads based on sunflower, canola spreads with omega-3 fatty acids, monounsaturated olive oil spreads, spreads with and without any dairy products, reduced fat spreads of all varieties and for those who read labels some spreads very low in trans fatty acids.

Recent introductions to the spreads market have included two brands containing plant sterols to aid in cholesterol management, a reduced fat butter and a spread based on a mixture of avocado and olive oils. All of these can be spread on toast or put in a sandwich. But what else can they be used for? Can they be used in baking? If so what sort of baking? Does this represent an opportunity to produce baked goods containing less calories from fat or do recipes need to be modified to compensate for the reduced fat? If a fat is to be classed as a margarine under FSANZ it must consist of at least 80% fat. If it has less than this amount it is classed as a spread. Most spreads are 70% or 75% fat. Table below shows typical fat contents of butter, margarine and spreads. It can be noted that the amount of salt varies considerably.

Table: *Comparison between Butter, Margarine (Polyunsaturated), Spreads and "Light" Products*

	% Fat	*% Moisture*	*% Salt*
Butter	81.5	16	2.4 – 2.6
Polyunsaturated Margarine	80	16	1.0 – 2.0
Typical Spread	70 – 75	20 – 25	1.0 – 1.8
Light Spread	60	35 –40	1.0
Low Fat Spread	40	60	1.0

All butter, margarine and spreads are water in oil emulsions. Margarine and spreads contain added emulsifiers such as lecithin and monoglycerides to aid in the emulsion preparation. Butter in contrast contains milkfat lecithin, a natural emulsifier.

Spreads were originally designed for table use and not specifically for baking. A consideration of the function of fats in baking should help us to decide which spreads to use for baking. Fat plays a different role in each baking application. For this reason in commercial practice a different formulation would be used for each purpose. But at home you do not really want to have three or four products in the refrigerator – two may suffice.

Bread Making

In bread making fat provides flavour but more importantly lubricates the dough. This helps to retain the gases released during baking thus ensuring a well risen loaf which will have a soft crumb and will stay fresh longer.

Because bread dough contains a considerable amount of water the water content of the fat used should not matter. However to achieve optimum results with a reduced fat spread more of the product should be used and the water added should be adjusted accordingly. In practice when using a home bread maker it has been found that some reduced fat spreads are not readily incorporated into the bread dough. This is because the spreads are made with a tight emulsion to bind the high amounts of water with the reduced fat. For the same reason reduced fat spreads on toast do not melt in the same way as butter.

Cake Making

The function of fat in cakes is more complex. Cakes are usually made by either the creaming method or the all in one method. The role of fat is different in each case. In the creaming method the fat is beaten with the sugar until it becomes light, fluffy and pale in colour. The water in oil emulsion changes to an oil in water emulsion. During this process air is being incorporated into the batter and the volume increases. The air in the batter is important because it forms nuclei or sites into which other gases, water vapour from the moisture and carbon dioxide from the baking powder, migrate and expand on heating.

It is important that the batter is stable enough to retain these gases and produce a cake with a light texture. The solids profile and crystal size of the fat have a major influence on aeration performance.

Even butters having the same fat content but sourced from different countries will produce different results. If a spread has good creaming properties with sugar and is able to trap and retain air then it should give an acceptable cake.

However the performance of each spread will depend on its solids profile and its crystalline structure. The only way for the home baker to assess this is by experimentation.

Fats containing 70% – 75% fat should give reasonable results. If they are substituted gram for gram in a recipe a cake would result containing approximately 10% less fat than the same cake made with traditional butter. The home baker might achieve a better cake if a small calculation were made and 10% more spread were added. A digital weighing scale would make this easy.

If using a reduced fat spread e.g. 60% fat the same reasoning would apply but now 25% more spread would be needed. This, however, is theoretical. The creaming process involves a dynamic balance of fat, sugar and water. A reduced fat spread with its higher proportion of water might upset this balance with unfavourable results.

Cakes or muffins made by the all in one method generally mix together all the liquid ingredients with baking soda in one bowl and all the dry ingredients including baking powder in another bowl. The two are then combined together before baking. In this method the rising agents are the baking soda and the baking powder.

The fat has no aeration function but as in bread making it will help to retain the gases released during baking. This will provide a good texture to the finished cake. In order to achieve the optimum result the amount of spread used should be adjusted to give the same amount of fat as in the original recipe. The amount of water or milk used should also be adjusted. i.e. reduce the water in the recipe by however many extra grams of spread are added.

Biscuits and Pastry

The lower the fat content of a spread the greater the proportion of water. What will be the effect of this water in biscuit and pastry making? Biscuits and pastry are made with "shortening ". The fat is rubbed into the flour. The molecules of fat surround the flour particles and exclude water. This prevents the development of gluten in the dough. The fat is said to shorten the dough. Any increase in water in the mixture will tend to encourage development of gluten, which will make biscuits hard and pastry heavy. Commercial bakers use different formulations for short pastry, Danish pastry and puff pastry.

A 75% spread could be used to make shortcrust pastry or for biscuits but none of the reduced fat or low fat spreads are suitable for Danish or puff pastry.

In the past, there were only all purpose margarines or fats for the bakery applications. For these applications not only the solid content of the fat (N-value) but also the consistency value (hardness) is important for a number of applications

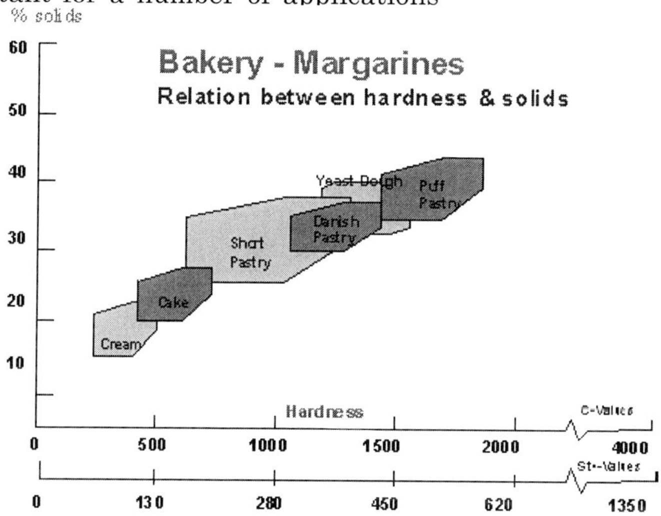

Figure: *This figure shows the relation between the solids and the hardness for various types of bakery margarines ($C/N^2=1$). It is indicating that at a temperature of application of 20°C, the N20 of the pastry margarine should be about 40 in order to get the required consistency (C-value of about 1600 or St4.4 value of about 500). This means also that a pastry margarine, which will be used at 30°C, should have a similar hardness at 30°C, resulting in a N30 of about 40.*

The minimal C20-value (hardness at 20°C) for a cream margarine should be about 300-400 in order to be able to pack as block of margarine; this means that the minimum N20 is about 20. If however a cream margarine is packed in a so called bag-in-box (liquid filled system) then the minimum solids at the critical distribution temperature can be lower like 13-15 %.

Margarine

Margarine is an imitation butter spread used for spreading, baking, and cooking. It was originally created from beef tallow and skimmed milk in 1869 in France by Hippolyte Mège-Mouriès, as a result of a challenge proposed by Emperor Louis Napoleon III, as a substitute for butter which would later be renamed "margarine".

Whereas butter is made from the butterfat of milk, modern margarine is made mainly of refined vegetable oil and water, and may also contain milk. In some locales it is colloquially referred to as "oleo", short for oleomargarine.

Margarine, like butter, consists of a water-in-fat emulsion, with tiny droplets of water dispersed uniformly throughout a fat phase which is in a stable crystalline form. Margarine has a minimum fat content of 80%, the same as butter, but unlike butter, reduced-fat varieties of margarine can also be labelled as margarine (in some countries). Colloquially in the US, the term margarine is used to describe "non-dairy spreads" like Country Crock, and I Can't Believe It's Not Butter! with varying amounts of fat content.

Margarine can be used for spreading, baking, and cooking. It is also commonly used as an ingredient in other food products, such as pastries and cookies, for its wide range of functionalities.

The principal raw material in the original formulation of margarine was beef fat. In 1871, Henry W. Bradley of Binghamton, New York patented a process for creating margarine that combined vegetable oils (primarily cottonseed oil) with animal fats. Shortages in beef fat supply combined with advances by Boyce and Sabatier in the hydrogenation of plant materials soon accelerated the use of Bradley's method, and between 1900 and 1920 commercial oleomargarine was produced from a combination of animal fats and hardened and unhardened vegetable oils. The depression of the 1930s, followed by the rationing of World War II, led to a reduction in supply of animal fat; and, by 1945, "original" margarine almost completely disappeared from the market. In the US, problems with supply, coupled with changes in legislation, caused manufacturers to switch almost completely to vegetable oils and fats (oleomargarine) by 1950 and the industry was ready for an era of product development.

While butter that cows produced had a slightly yellow colour, margarine had a white colour, making the margarine look more like lard. Many people found it to look unappetizing. Around the late 1880s the manufacturers decided to dye the margarine yellow, so it would sell more. Dairy firms, especially in Wisconsin, became alarmed and succeeded in getting legislation passed to prohibit the colouring of the stark white product. In response, the margarine companies distributed the margarine together with a packet of yellow dye. The product was placed in a bowl and the dye mixed in with a spoon. This took some time and effort and it was not unusual for the final product to be served as a light and dark yellow, or even white, striped product. During World War II, there was a shortage of butter in the United States and "oleomargarine" became popular. In 1951 the W. E. Dennison Company received patent number 2,553,513 for a method to place a capsule of yellow dye inside a plastic package of margarine. After purchase, the capsule was broken inside the package and then the package was kneaded to distribute the dye. Around 1955, the artificial colouring laws were repealed and margarine could for the first time be sold coloured like butter.

During WWII rationing, only two types of margarine were available in the UK, a premium brand and a cheaper budget brand. With the end of rationing in 1955, the market was opened to the forces of supply and demand, and brand marketing became prevalent. The competition between the major producers was given further impetus with the beginning of commercial television advertising in 1955 and, throughout the 1950s and 1960s, competing companies vied with each other to produce the margarine that tasted most like butter.

In the mid-1960s, the introduction of two lower-fat blends of butter oil and vegetable oils in Scandinavia, called *Lätt & Lagom* and *Bregott*, clouded the issue of what should be called "margarine" and began the debate that led to the introduction of the term "spread". In 1978, an 80% fat product called *krona*, made by churning a blend of dairy cream and vegetable oils, was introduced in Europe and, in 1982, a blend of cream and vegetable oils called *clover* was introduced in the UK by the Milk Marketing Board. The vegetable oil and cream spread I Can't Believe It's Not Butter! was introduced into the United States in 1981 and in the United Kingdom and Canada in 1991.

In recent decades, margarine spreads have gone through many developments in efforts to improve their healthfulness. Most brands have phased out the use of hydrogenated oils, and are now also trans fat free. Many brands have launched refrigerator-stable margarine spreads that contain only 1/3 of the fat and calorie content of traditional spreads. Other varieties of spreads include those with added Omega-3 fatty acids, those with low or no salt, those with added plant sterols, claimed to reduce blood cholesterol, and some made from olive oil or certified vegan oils.

Manufacturing Process

The basic method of making margarine today consists of emulsifying a blend of vegetable oils and fats, which can be modified using fractionation, interesterification, and/or hydrogenation, with skimmed milk, chilling the mixture to solidify it and working it to improve the texture. Vegetable and animal fats are similar compounds with different melting points. Those fats that are liquid at room temperature are generally known as oils. The melting points are related to the presence of carbon-carbon double bonds in the fatty acids components. Higher number of double bonds give lower melting points.

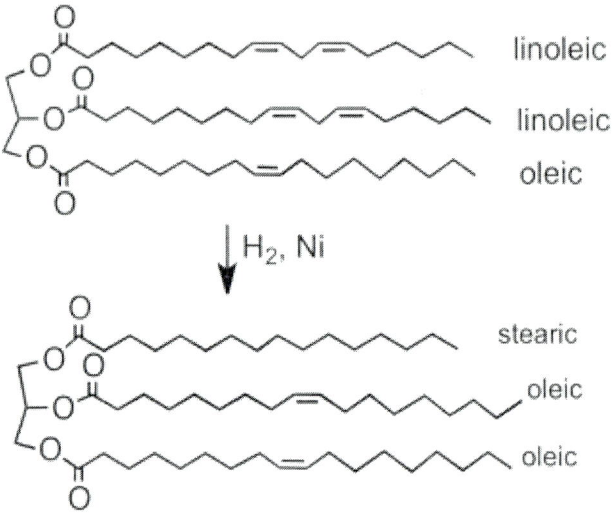

Figure: *Partial hydrogenation of a typical plant oil to a typical component of margarine. Most of the C=C double bonds are removed in this process, which elevates the melting point of the product.*

Commonly, the natural oils are hydrogenated by passing hydrogen through the oil in the presence of a nickel catalyst, under controlled conditions. The addition of hydrogen to the unsaturated bonds (alkenic double C=C bonds) results in saturated C-C bonds, effectively increasing the melting point of the oil and thus "hardening" it. This is due to the increase in van der Waals' forces between the saturated molecules compared with the unsaturated molecules. However, as there are possible health benefits in limiting the amount of saturated fats in the human diet, the process is controlled so that only enough of the bonds are hydrogenated to give the required texture. Margarines made in this way are said to contain hydrogenated fat. This method is used today for some margarines although the process has been developed and sometimes other metal catalysts are used such as palladium. If hydrogenation is incomplete (partial hardening), the relatively high temperatures used in the hydrogenation process tend to flip some of the carbon-carbon double bonds into the "trans" form.

If these particular bonds aren't hydrogenated during the process, they will still be present in the final margarine in molecules of trans fats, the consumption of which has been shown to be a risk factor for cardiovascular disease. For this reason, partially hardened fats are used less and less in the margarine industry. Some tropical oils, such as palm oil and coconut oil, are naturally semi solid and do not require hydrogenation.

Modern margarines can be made from any of a wide variety of animal or vegetable fats, mixed with skim milk, salt, and emulsifiers. Margarines and vegetable fat spreads found in the market can range from 10 to 90% fat. Depending on its final fat content and its purpose (spreading, cooking or baking), the level of water and the vegetable oils used will slightly vary. The oil is pressed from seeds and refined. It is then blended with solid fat. If no solid fats are added to the vegetable oils, the latter undergo a full or partial hydrogenation process to solidify them. The resulting blend is mixed with water, citric acid, carotenoids, vitamins and milk powder. Emulsifiers such as lecithin help disperse the water phase evenly throughout the oil, and salt and preservatives are also commonly added. This oil and water emulsion is then heated, blended, and cooled. The softer tub margarines are made with less hydrogenated, more liquid, oils than block margarines.

Three types of margarine are common:

- Soft vegetable fat spreads, high in mono- or polyunsaturated fats, which are made from safflower, sunflower, soybean, cottonseed, rapeseed, or olive oil.
- Margarines in bottle to cook or top dishes
- Hard, generally uncoloured margarine for cooking or baking. (Shortening)

Blending with Butter

Many popular table spreads sold today are blends of margarine and butter or other milk products. Blending, which is used to improve the taste of margarine, was long illegal in countries such as the United States and Australia. Under European Union directives, a margarine product cannot be called "butter", even if consists mostly of natural butter. In some European countries butter-based table spreads and margarine products are marketed as "butter mixtures".

Butter mixtures now make up a significant portion of the table spread market. The brand "I Can't Believe It's Not Butter!" spawned a variety of similarly named spreads that can now be found on supermarket shelves all over the world, with names like "Beautifully Butterfully", "Butterlicious", "Utterly Butterly", and "You'd Butter Believe It". These butter mixtures avoid the restrictions on labelling, with marketing techniques that imply a strong similarity to real butter. Such marketable names present the product to consumers differently from the required product labels that call margarine "partially hydrogenated vegetable oil".

Nutrition

Discussions concerning the nutritional value of margarines and spreads revolve around two aspects — the total amount of fat, and the types of fat (saturated fat, trans fat). Usually, a comparison between margarine and butter is included in this context as well.

Amount of Fat

The roles of butter and traditional margarine (80% fat) are similar with respect to their energy content, but low-fat margarines and spreads are also widely available.

Saturated Fat

Saturated fatty acids have not been conclusively linked to elevated blood cholesterol levels. Replacing saturated and trans unsaturated fats with unhydrogenated monounsaturated and polyunsaturated fats is more effective in preventing coronary heart disease in women than reducing overall fat intake.

Vegetable fats can contain anything from 7% and 86% saturated fatty acids. Liquid oils (canola oil, sunflower oil) tend to be on the low end, while tropical oils (coconut oil, palm kernel oil) and fully hardened (hydrogenated) oils are at the high end of the scale. A margarine blend is a mixture of both types of components. Generally, firmer margarines contain more saturated fat.

Typical soft tub margarine contains 10% to 20% of saturated fat. Regular butterfat contains 52 to 65% saturated fats.

Unsaturated Fat

Consumption of unsaturated fatty acids has been found to decrease LDL cholesterol levels and increase HDL cholesterol levels in the blood, thus reducing the risk of contracting cardiovascular diseases.

There are two types of unsaturated oils: mono- and poly-unsaturated fats both of which are recognized as beneficial to health in contrast to saturated fats. Some widely grown vegetable oils, such as rapeseed (and its variant canola), sunflower, safflower, and olive oils contain high amounts of unsaturated fats. During the manufacture of margarine, some of the unsaturated fats may be converted into hydrogenated fats or trans fats in order to give them a higher melting point so that they are solid at room temperatures.

Omega-3 Fatty Acids

Omega-3 fatty acids are a family of polyunsaturated fatty acids, which many claim have been found especially good for health. This is one of the two Essential fatty acids, so called because humans cannot manufacture it and must get it from food. Omega-3 fatty acids are mostly obtained from oily fish caught in high-latitude waters. They are comparatively uncommon in vegetable sources, including margarine. However, one type of Omega-3 fatty acid, alpha-Linolenic acid (ALA) can be found in some vegetable oils. Flax oil contains 30-50% of ALA, and is becoming a popular dietary supplement to rival fish oils; both are often added to premium margarines. An ancient oil plant, *camelina sativa*, has recently gained popularity because of its high Omega-3 content (30-45%), and it has been added to some margarines. Hemp oil contains about 20% ALA. Small amounts of ALA are found in vegetable oils such as soybean oil (7%), rapeseed oil (7%) and wheat germ oil (5%).

Omega-6 Fatty Acids

Omega-6 fatty acids are also important for health. They include the essential fatty acid linoleic acid (LA), which is abundant in vegetable oils grown in temperate climates. Some, such as hemp (60%) and the common margarine oils corn (60%), cottonseed (50%) and sunflower (50%), have large amounts, but most temperate oil seeds have over 10% LA. Margarine is very high in omega-6 fatty acids. Modern Western diets are frequently quite high in Omega-6 but very deficient in Omega-3. The omega-6 to omega-3 ratio is typically 5:1 to 10:1. Large amounts of omega-6 decreases the effect of omega-3. Therefore it is recommended that the ratio in the diet should be less than 4:1, although optimal ratio may be closer to 1:1.

Trans Fat

Unlike essential fatty acids, saturated and trans fatty acids are not essential and provide no known benefit to human health besides providing calories. There is a positive linear trend between saturated and trans fatty acid intake and LDL cholesterol concentration, and therefore increased risk of coronary heart disease, by raising levels of LDL cholesterol and lowering levels of HDL cholesterol.

Several large studies have indicated a link between consumption of high amounts of trans fat and coronary heart disease, and possibly some other diseases, prompting a number of government health agencies across the world to recommend that the intake of trans-fats be minimized. In the US, partial hydrogenation has been common as a result of preference for domestically produced oils. However, since the mid-1990s, many countries around the world have started to move away from using partially hydrogenated oils. This led to the production of new margarine varieties that contain less or no trans fat.

Cholesterols

High levels of cholesterol, particularly of the LDL kind is associated with an increased risk of atherosclerosis and atheroma formation. These narrowing of important blood vessels can cause reduced blood flow to the brain, heart, kidneys and other parts of the body. Cholesterol, though needed metabolically, is not essential in the diet. The human body makes cholesterol in the liver, adapting the production according to its food intake, producing about 1g of cholesterol each day or 80% of the needed total body cholesterol. The remaining 20% comes directly from food intake.

Therefore overall intake of cholesterol as food has less effect on blood cholesterol levels than the type of fat eaten. However, some individuals are more responsive to dietary cholesterol than others. The US Food and Drug Administration states that healthy people should not consume more than 300 mg of cholesterol each day. Most margarines are vegetable-based and thus contain no cholesterol, while 100 grams of butter contains 178 mg of cholesterol.

Plant Sterol Esters and Stanol Esters

Plant sterol esters or plant stanol esters have been added to some margarines and spreads because of their cholesterol lowering effect. Several studies have indicated that consumption of about 2 grams per day provides a reduction in LDL cholesterol of about 10%.

Market Acceptance

Margarine, particularly polyunsaturated margarine, has become a major part of the Western diet and had overtaken butter in popularity in the mid-20th century. In the United States, for example, in 1930, the average person ate over 18 lb (8.2 kg) of butter a year and just over 2 lb (0.91 kg) of margarine. By the end of the 20th century, an average American ate around 5 lb (2.3 kg) of butter and nearly 8 lb (3.6 kg) of margarine.

Margarine has a particular market value to those who observe the Jewish dietary laws of Kashrut. Kashrut forbids the mixing of meat and dairy products; hence there are strictly Kosher non-dairy margarines available.

These are often used by the Kosher consumer to adapt recipes that use meat and butter or in baked goods that will be served with meat meals. The 2008 Passover margarine shortage in America caused much consternation within the Kosher-observant community.

Regular margarine contains trace amounts of animal products such as whey or dairy casein extracts. However, margarine that strictly doesn't contain animal products also exists. Such margarines provide a vegan substitute for butter.

National Standards

Australia

Margarine is common in Australian supermarkets. Sales of the product have decreased in recent years due to consumers "reducing their use of spreads in their daily diet".

Butter-coloured margarine was sold from its introduction in Australia, but dairy and associated industries lobbied governments strongly in a (vain) attempt to have them change its colour, or banned altogether.

Australia New Zealand Food Standards Code - Standard 2.4.2 - Edible Oil Spreads requires that edible oil spreads and table margarine must contain no less than 55 μg/kg of vitamin D.

Canada

Canadian standard B.09.016 states that margarine shall be: "An emulsion of fat, or water in fat, oil, or fat and oil that are not derived from milk and shall contain not less than 80% fat and not less than 3300 IU of vitamin A and 530 IU of vitamin D."

Calorie reduced margarine is specified in standard B.09.017 as: "Containing not less than 40% fat and having 50% of the calories normally present in margarine."

In 2007, Health Canada released an updated version of the Canada's Food Guide which recommended Canadians choose "soft" margarine spreads that are low in saturated and trans fats and limit the use of traditional "hard" margarines, butter, lard, and shortening in their diets.

European Union

Under European Union directives, margarine is defined as: "A water-in-oil emulsion derived from vegetable/animal fats, with a fat content of at least 80% but less than 90%, that remain solid at a temperature of 20°C and are suitable as spread."

Margarines may not have a milk fat content of more than 3%. For blends and blended spreads, the milk fat may be between 10% and 80%.

Spread that contains 60 to 62% of fat may be called "three-quarter-fat margarine" or "reduced-fat margarine". Spread that contains 39 to 41% of fat may be called "half-fat margarine", "low-fat margarine", or "light margarine". Spreads with any other percentage of fat are called "fat spread" or "light spread".

Many member states currently require the mandatory addition of vitamins A and D to margarine and fat spreads for reasons of public health. Voluntary fortification of margarine with vitamins had been practiced by manufacturers since 1925, but in 1940 with the advent of the war, certain governments took action to safeguard the nutritional status of their nations by making the addition of vitamin A and D compulsory. This mandatory fortification was justified in the view that margarine was being used to replace butter in the diet.

United Kingdom

In the United Kingdom there are no brands of spread on sale which contain any partially hydrogenated oils. Fortification with vitamins A and D is no longer mandatory for margarine, this brings it in line with other spreads where it is only a voluntary requirement.

Legal Issues

Since margarine intrinsically appears white or almost white, by forbidding the addition of artificial colouring agents, legislators found they could protect the dairy industries by discouraging the consumption of margarine based on visual appeal (if margarine were coloured the same as butter, consumers would see it as being virtually the same thing as butter, and as a natural product, even though it is not). Bans on adding colour became commonplace in the US, Australasia, Canada, and Denmark and, in some cases, those bans endured for almost 100 years. It did not become legal to sell coloured margarine in Australia, for example, until the 1960s.

The rivalry between the dairy industry and the oleomargarine industry persists even today

Shortening

Shortening is any fat that is solid at room temperature and used to make crumbly pastry. Shortening is used in pastries that should not be elastic, such as cake. Although butter is solid at room temperature and is frequently used in making pastry, the term "shortening" seldom refers to butter, but is more closely related to margarine.

History and Market

Originally shortening was synonymous with lard, but with the invention of margarine by French chemist Hippolyte Mège-Mouriès in 1869, margarine also came to be included in the term. Since the invention of hydrogenated vegetable oil in the early 20th century, "shortening" has come almost exclusively to mean hydrogenated vegetable oil. Vegetable shortening shares many properties with lard: Both are semi-solid fats with a higher smoke point than butter and margarine. They contain less water and are thus less prone to splattering, making them safer for frying.

Lard and shortening have a higher fat content compared to about 80% for butter and margarine. Cake margarines and shortenings tend to contain a few percent of monoglycerides whereas margarines typically have less. Such "high ratio shortenings" blend better with hydrophilic ingredients such as starches and sugar.

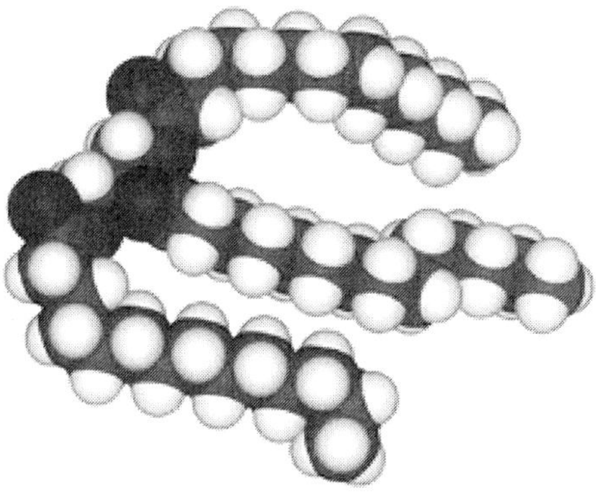

Figure: *A triglyceride molecule, the main constituent of shortening.*

Hydrogenation of organic substances was first developed by the French chemist Paul Sabatier in 1897, and in 1901 the German chemist Wilhelm Normann developed the hydrogenation of fats, which he patented in 1902. In 1907, a German chemist, Edwin Cuno Kayser, moved to Cincinnati, Ohio, the home town of soap manufacturer Procter & Gamble. He had worked for British soap manufacturer

Joseph Crosfield and Sons and was well acquainted with Normann's process, as Crosfield and Sons owned the British rights to Normann's patent. Soon after arrival, Kayser made a business deal with Procter & Gamble, and presented the company with two processes to hydrogenate cottonseed oil, with the intent of creating a raw material for soap.

Since the product looked like lard, Procter & Gamble instead began selling it as a vegetable fat for cooking purposes in June 1911, calling it "Crisco", a modification of the phrase "crystallized cottonseed oil". While similar to lard, vegetable shortening was much cheaper to produce. Shortening also required no refrigeration, which further lowered its costs and increased its appeal in a time when refrigerators were rare. With these advantages, plus an intensive advertisement campaign by Procter & Gamble, Crisco quickly gained popularity in American households.

As food production became increasingly industrialized and manufacturers sought low-cost raw materials, the use of vegetable shortening also became common in the food industry. In addition, vast US government-financed surpluses of cottonseed oil, corn oil, and soy beans also helped create a market in low-cost vegetable shortening.

Crisco, owned by The J.M. Smucker Company since 2002, remains the best-known brand of shortening in the US, nowadays consisting of a blend of partially and fully hydrogenated soybean and palm oils. In Ireland and the UK, Cookeen is a popular brand, while in Australia, Copha is popular, although made primarily from coconut oil.

Shortened Dough

A short dough is one that is crumbly or mealy. The opposite of a short dough is a "long" dough or dough that stretches. Vegetable shortening (or butter, or other solid fats) can produce both types of dough; the difference is in technique. To produce a short dough, which is commonly used for tarts, the shortening is cut into the flour with a pastry blender, pair of table knives, fingers, or other utensil until the resulting mixture has a fine, cornmeal-like texture.

For a long dough, the shortening is cut in only until the pea-sized crumbs are formed, or even larger lumps may be included. After cutting in the fat, the liquid (if any) is added and the dough is shaped for baking.

Neither short dough nor long flake dough are considered to be either creamed or stirred batters.

Plastic Bakery Fats

There are two types of fat used in baking, SOLID FAT, such as stick butter or margarine or shortening, or LIQUID, such as vegetable or olive oil. These types are extracted from either plants or animals, or manufactured. There are other ingredients that contribute fat to baked goods such as egg yolks, milk, chocolate, nuts and seeds, etc, but are discussed in their separate ingredient sections.

Solid fat, as in butter or shortening, is technically referred to as "plastic" fat because when beaten with sugar, such as in a buttercake recipe, they can hold air bubbles in their creamy, malleable mass. These types can also be used as a spacer, such as a pie crust or for leavening, such as in puff pastry.

Liquid fats, such as oil have different characteristics than solid ones and cannot hold air, and can be used in small amounts for healthy recipes.

Fats Used For Baking

TYPE	*FAT CONTENT - remaining percent is water, milk solids, etc.*	*FLAVOR - Butter has the most*	*Cakes - tender or not*	*Cookies - spread or not*	*Pies - flaky or not*
SOLID "PLASTIC" FAT					
Stick Butter	81% fat	Yes	Medium	Some	Medium
Lard	100%	No	Most	Least	Most
Stick Margarine	80%	Some	Least	Most	Least
Shortening	100%	No	Most	Least	Most
"LIQUID" FATS					
Vegetable Oil	100%	None	Dense	Cakelike	None
Olive Oil	100%	Some	Dense	Cakelike	None
Clarified Butter	100%	Some	Dense	Cakelike	None

Sarah Says: I have done a lot of testing recipes with different types fats because of my developing hundreds of low-fat and fat-reduced recipes. Here, I share my first-hand knowledge and experiences with you, which may contradict other websites. But, my findings here have been seen with my own eyes!! Happy Baking, Sarah

Please Note that when substituting a solid fat, such as butter with a liquid one (vegetable oil) or vice versa, adversely affects the taste or the texture of your recipe.

Melted butter cannot be substituted with a liquid fat because butter contains milk solids which set after the recipe has baked and cooled; oil doesn't. The rule of thumb is to substitute one solid fat for another (ie: butter for shortening), and oil with another oil (vegetable oil for canola oil). That's because when a fat is solid, it acts differently in a baking recipe than a fat that's liquid.

If a recipe calls for butter, plastic (solid) fats such as butter, margarine or shortening with at least a 80% fat content work the best. Fats containing less than 80% in fat will adversely affect your recipe, such as lower calorie "spreads". Avoid their use.

Pastry

Pastry is a major type of bakers' confectionery. It includes many of the various kinds of baked products made from ingredients such as flour, sugar, milk, butter, shortening, baking powder, and eggs. Small tarts and other sweet baked products are called pastries. Common pastry dishes include pies, tarts, quiches and pasties.

Pastry can also refer to the pastry dough, from which such baked products are made. Pastry dough is rolled out thinly and used as a base for baked products.

Pastry is differentiated from bread by having a higher fat content, which contributes to a flaky or crumbly texture. A good pastry is light and airy and fatty, but firm enough to support the weight of the filling. When making a shortcrust pastry, care must be taken to blend the fat and flour thoroughly before adding any liquid. This ensures that the flour granules are adequately coated with fat and less likely to develop gluten. On the other hand, overmixing results in long gluten strands that toughen the pastry. In other types of pastry such as Danish pastry and croissants, the characteristic flaky texture is achieved by repeatedly rolling out a dough similar to that for yeast bread, spreading it with butter, and folding it to produce many thin layers.

Shortcrust Pastry

Shortcrust pastry is the simplest and most common pastry. It is made with flour, fat, butter, salt, and water to bind the dough. This is used mainly in tarts. It is also the pastry that is used most often in making a quiche.

The process of making pastry includes mixing of the fat and flour, adding water, and rolling out the paste. The fat is mixed with the flour first, generally by rubbing with fingers or a pastry blender, which inhibits gluten formation by coating the gluten strands in fat and results in a short (as in crumbly; hence the term shortcrust), tender pastry.

A related type is the sweetened sweetcrust pastry, also known as *paté sucrée*, in which sugar and egg yolks have been added (rather than water) to bind the pastry

Chemistry

Different kinds of pastries are made by utilizing the natural characteristics of wheat flour and certain fats. When wheat flour is mixed with water and kneaded into plain dough, it develops strands of gluten, which are what make bread tough and elastic.

In a typical pastry, however, this toughness is unwanted, so fat or oil is added to slow down the development of gluten. Lard or suet work well because they have a coarse, crystalline structure that is very effective. Using unclarified butter does not work well because of its water content; clarified butter, which is virtually water-free, is better, but shortcrust pastry using only butter may develop an inferior texture. If the fat is melted with hot water or if liquid oil is used, the thin oily layer between the grains offers less of an obstacle to gluten formation and the resulting pastry is tougher.

Puff Pastry

In baking, a puff pastry, also referred to as "pβte feuilletée," is a light, flaky, mechanically leavened pastry containing several layers of fat which is in solid state at 20 °C (68 °F). In raw form, puff pastry is a laminated dough composed of two elements: a "dough packet", the détrempe and a "butter packet" or other solid fat, the beurrage. In classic puff pastry, an envelope is formed by placing the beurrage inside the détrempe.

In "inverse puff" pastry, an envelope is formed by placing the détrempe inside the beurrage. The resulting paton is repeatedly folded and rolled out.

The gaps that form between the layers left by the fat melting are pushed (leavened) by the water turning into steam during the baking process. Piercing the dough will prevent excessive puffing, and crimping along the sides will prevent the layers from flaking all of the way to the edges.

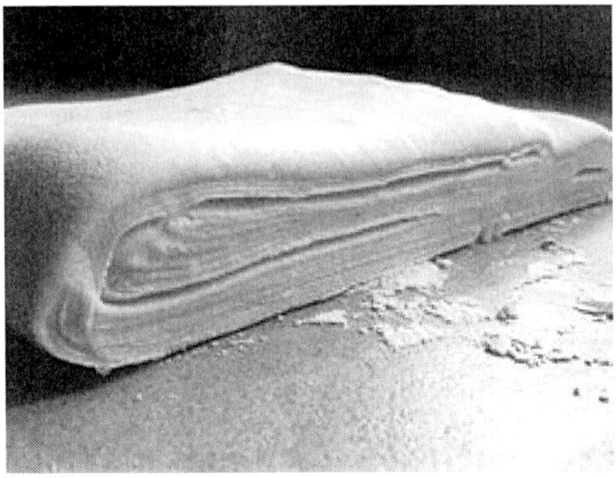

Figure: *Puff pastry before baking, with layers clearly visible*

Production

Figure: *A palmier, or "palm leaf", design*

The production of puff pastry dough can be time-consuming, because it must be kept at a temperature of approximately 16 °C (60 °F) to keep shortening from becoming runny, and must rest in between folds to allow gluten strands time to link up and thus retain layering.

The number of layers in puff pastry is calculated with the equation:

$$l = (f + 1)^n$$

where l is the number of finished layers, f the number of folds, and n the number of times the dough has been folded. For example, twice-folding (i.e. in three) for four times gives $(2+1)^4 = 81$ layers. Chef Julia Child recommends 73 layers for regular *pβte feuilletée* and 729 (i.e. 3^6) layers for *pβte feuilletée fine* (in Volume II of her *Mastering the Art of French Cooking* textbook).

Commercially made puff pastry is available in grocery stores. Common types of fat used include butter, vegetable shortenings, and lard. Butter is the most common type used because it provides a richer taste and superior mouthfeel.

Since shortenings and lard have a higher melting point, puff pastry made with either will rise more than pastry made with butter if made correctly; however it will often have a waxy mouthfeel and a blander flavour.

Specialized margarine formulated for high plasticity (the ability to spread very thin without breaking apart) is used for industrial production of puff pastry.

Puff pastry is not the same as phyllo (filo) pastry, although puff pastry can be substituted for phyllo in some applications. Phyllo dough is made with flour, water, and fat and is stretched to size rather than rolled. Usually when using phyllo dough, a small amount of oil or melted fat (usually butter) is brushed on one layer of phyllo dough and is topped with another layer.

This process can be repeated as many times as desired. When it bakes, it becomes crispy but, since it contains somewhat less water, does not expand to the same degree as puff pastry does.

Nor is puff pastry the same as Austrian strudel dough, or Strudelteig. Strudel dough is more like the phyllo described above.

Variants

Puff pastry can also be leavened with baker's yeast to create croissants or Danish pastry or Spanish/Portuguese milhoja and empanadilla, though such doughs are not universally known as puff pastries.

In addition, since the process of making puff pastry is generally somewhat laborious and quite time-intensive, faster recipes (known as "blitz", "rough puff", or "Flaky pastry") are fairly common. Many of these recipes combine the butter into the détrempe rather than adding it in the folding process and are thus similar to a folded short crust.

Cake

Cake is a form of bread or bread-like food. In its modern forms, it is typically a sweet baked dessert. In its oldest forms, cakes were normally fried breads or cheesecakes, and normally had a disk shape. Determining whether a given food should be classified as bread, cake, or pastry can be difficult.

Figure: *A layered pound cake filled with raspberry jam and lemon curd, and finished with buttercream frosting*

Modern cake, especially layer cakes, normally contain a combination of flour, sugar, eggs, and butter or oil, with some varieties also requiring liquid (typically milk or water) and leavening agents (such as yeast or baking powder). Flavourful ingredients like fruit purées, nuts, dried or candied fruit, or extracts are often added, and numerous substitutions for the primary ingredients are possible. Cakes are often filled with fruit preserves or dessert sauces (like pastry cream), iced with buttercream or other icings, and decorated with marzipan, piped borders, or candied fruit.

Cake is often the dessert of choice for meals at ceremonial occasions, particularly weddings, anniversaries, and birthdays. There are countless cake recipes; some are bread-like, some rich and elaborate, and many are centuries old. Cake making is no longer a complicated procedure; while at one time considerable labour went into cake making (particularly the whisking of egg foams), baking equipment and directions have been simplified so that even the most amateur cook may bake a cake.

Varieties

Cakes are broadly divided into several categories, based primarily on ingredients and cooking techniques.

- Yeast cakes are the oldest and are very similar to yeast breads. Such cakes are often very traditional in form, and include such pastries as babka and stollen.
- Cheesecakes, despite their name, aren't really cakes at all. Cheesecakes are in fact custard pies, with a filling made mostly of some form of cheese (often cream cheese, mascarpone, ricotta, or the like), and have very little flour added, although a flour-based or graham cracker crust may be used. Cheesecakes are

also very old, with evidence of honey-sweetened cakes dating back to ancient Greece.

- Sponge cakes are thought to be the first of the non-yeast-based cakes and rely primarily on trapped air in a protein matrix (generally of beaten eggs) to provide leavening, sometimes with a bit of baking powder or other chemical leaven added as insurance.

  Such cakes include the Italian/Jewish *pan di Spagna* and the French Génoise. Highly decorated sponge cakes with lavish toppings are sometimes called *gateau*; the French word for cake.

- Butter cakes, including the pound cake and devil's food cake, rely on the combination of butter, eggs, and sometimes baking powder or bicarbonate of soda to provide both lift and a moist texture.

Beyond these classifications, cakes can be classified based on their appropriate accompaniment (such as coffee cake) and contents (e.g. fruitcake or flourless chocolate cake).

Some varieties of cake are widely available in the form of cake mixes, wherein some of the ingredients (usually flour, sugar, flavouring, baking powder, and sometimes some form of fat) are premixed, and the cook needs add only a few extra ingredients, usually eggs, water, and sometimes vegetable oil or butter. While the diversity of represented styles is limited, cake mixes do provide an easy and readily available homemade option for cooks who are not accomplished bakers.

Special-purpose Cakes

Cakes may be classified according to the occasion for which they are intended. For example, wedding cakes, birthday cakes, cakes for first communion, Christmas cakes, Halloween cakes, and Passover plava (a type of sponge cake sometimes made with matzo meal) are all identified primarily according to the celebration they are intended to accompany. The cutting of a wedding cake constitutes a social ceremony in some cultures. The Ancient Roman marriage ritual of *confarreatio* originated in the sharing of a cake.

Particular types of cake may be associated with particular festivals, such as stollen or chocolate log (at Christmas), babka and simnel cake (at Easter), or mooncake. There has been a long tradition of decorating an iced cake at Christmas time; other cakes associated with Christmas include chocolate log and mince pies.

Figure: *Cake made for a baby shower and decorated with edible ingredients*

The Influence of Emulsifiers in Baking

An emulsion is a mixture of two or more liquids that are normally immiscible (unmixable or unblendable). Emulsions are part of a more general class of two-phase systems of matter called colloids. Although the terms *colloid* and *emulsion* are sometimes used interchangeably, *emulsion* should be used when both the dispersed and the continuous phase are liquids. In an emulsion, one liquid (the dispersed phase) is dispersed in the other (the continuous phase). Examples of emulsions include vinaigrettes, milk, mayonnaise, and some cutting fluids for metal working.

The word "emulsion" comes from the Latin word for "to milk", as milk is an emulsion of fat and water, among other components.

Two liquids can form different types of emulsions. As an example, oil and water can form, first, an oil-in-water emulsion, wherein the oil is the dispersed phase, and water is the dispersion medium. Second, they can form a water-in-oil emulsion, wherein water is the dispersed phase and oil is the external phase. Multiple emulsions are also possible, including a "water-in-oil-in-water" emulsion and an "oil-in-water-in-oil" emulsion.

Emulsions, being liquids, do not exhibit a static internal structure. The droplets dispersed in the liquid matrix (called the "dispersion medium") are usually assumed to be statistically distributed. The term "emulsion" is also used to refer to the photo-sensitive side of photographic film. Such a photographic emulsion consist of silver halide colloidal particles dispersed in a gelatin matrix. Nuclear emulsions are similar to photographic emulsions, but used in particle physics to detect high-energy elementary particles.

Appearance and Properties

Emulsions contain both a dispersed and a continuous phase, with the boundary between the phases called the "interface". Emulsions tend to have a cloudy appearance because the many phase interfaces scatter light as it passes through the emulsion. Emulsions appear white when all light is scattered equally. If the emulsion is dilute enough, higher–frequency and low-wavelength light will be scattered more, and the emulsion will appear bluer – this is called the "Tyndall effect". If the emulsion is concentrated enough, the colour will be distorted towards comparatively longer wavelengths, and will appear more yellow. This phenomenon is easily observable when comparing skimmed milk, which contains little fat, to cream, which contains a much higher concentration of milk fat. One example would be a mixture of water and oil.

Two special classes of emulsions – microemulsions and nanoemulsions, with droplet sizes below 100 nm – appear translucent. This property is due to the fact that lightwaves are scattered by the droplets only if their sizes exceed about one-quarter of the wavelength of the incident light. Since the visible spectrum of light is composed of wavelengths between 390 and 750 nanometres (nm), if the droplet sizes in the emulsion are below about 100 nm, the light can penetrate through the emulsion without being scattered.

Due to their similarity in appearance, translucent nanoemulsions and microemulsions are frequently confused. Unlike translucent nanoemulsions, which require specialized equipment to be produced, microemulsions are spontaneously formed by "solubilizing" oil molecules with a mixture of surfactants, co-surfactants, and co-solvents. The required surfactant concentration in a microemulsion is, however, several times higher than that in a translucent nanoemulsion, and significantly exceeds the concentration of the dispersed phase. Because of many undesirable side-effects caused by surfactants, their presence is disadvantageous or prohibitive in many applications. In addition, the stability of a microemulsion is often easily compromised by dilution, by heating, or by changing pH levels.

Common emulsions are inherently unstable and, thus, do not tend to form spontaneously. Energy input – through shaking, stirring, homogenizing, or exposure to power ultrasound – is needed to form an emulsion. Over time, emulsions tend to revert to the stable state of the phases comprising the emulsion.

An example of this is seen in the separation of the oil and vinegar components of vinaigrette, an unstable emulsion that will quickly separate unless shaken almost continuously. There are important exceptions to this rule – microemulsions are thermodynamically stable, while translucent nanoemulsions are kinetically stable.

Whether an emulsion of oil and water turns into a "water-in-oil" emulsion or an "oil-in-water" emulsion depends on the volume fraction of both phases and the type of emulsifier (surfactant) present. In general, the Bancroft rule applies. Emulsifiers and emulsifying particles tend to promote dispersion of the phase in which they do not dissolve very well.

For example, proteins dissolve better in water than in oil, and so tend to form oil-in-water emulsions (that is, they promote the dispersion of oil droplets throughout a continuous phase of water).

The geometric structure of an emulsion mixture of two lyophobic liquids with a large concentration of the secondary component is fractal: Emulsion particles unavoidably form dynamic inhomogeneous structures on small length scale. The geometry of these structures is fractal. The size of elementary irregularities is governed by a universal function which depends on the volume content of the components. The fractal dimension of these irregularities is 2.5.

Instability

Emulsion stability refers to the ability of an emulsion to resist change in its properties over time. There are four types of instability in emulsions: flocculation, creaming, coalescence, and Ostwald ripening. Flocculation occurs when there is an attractive force between the droplets, so they form flocs, like bunches of grapes. Coalescence occurs when droplets bump into each other and combine to form a larger droplet, so the average droplet size increases over time. Emulsions can also undergo creaming, where the droplets rise to the top of the emulsion under the influence of buoyancy, or under the influence of the centripetal force induced when a centrifuge is used.

An appropriate "surface active agent" (or "surfactant") can increase the kinetic stability of an emulsion so that the size of the droplets does not change significantly with time. It is then said to be stable.

Monitoring Physical Stability

The stability of emulsions can be characterized using techniques such as light scattering, focused beam reflectance measurement, centrifugation, and rheology. Each method has advantages and disadvantages.

Accelerating Methods for Shelf Life Prediction

The kinetic process of destabilization can be rather long – up to several months, or even years for some products. Often the formulator must accelerate this process in order to test products in a reasonable time during product design. Thermal methods are the most commonly used - these consist of increasing the emulsion temperature to accelerate destabilization (if below critical temperatures for phase inversion or chemical degradation).

Temperature affects not only the viscosity but also the interfacial tension in the case of non-ionic surfactants or, on a broader scope, interactions of forces inside the system. Storing an emulsion at high temperatures enables the simulation of realistic conditions for a product (e.g., a tube of sunscreen emulsion in a car in the summer heat), but also to accelerate destabilization processes up to 200 times. Mechanical methods of acceleration, including vibration, centrifugation, and agitation, can also be used. These methods are almost always empirical, without a sound scientific basis.

Emulsifiers

An emulsifier (also known as an "emulgent") is a substance that stabilizes an emulsion by increasing its kinetic stability. One class of emulsifiers is known as "surface active agents", or surfactants.

Examples of food emulsifiers are:

- Egg yolk – in which the main emulsifying agent is lecithin. In fact, *lecithos* is the Greek word for egg yolk.
- Mustard – where a variety of chemicals in the mucilage surrounding the seed hull act as emulsifiers
- Soy lecithin is another emulsifier and thickener
- Pickering stabilization – uses particles under certain circumstances
- Sodium phosphates
- Sodium stearoyl lactylate
- DATEM (Diacetyl Tartaric (Acid) Ester of Monoglyceride) – an emulsifier used primarily in baking

Detergents are another class of surfactants, and will physically interact with both oil and water, thus stabilizing the interface between the oil and water droplets in suspension. This principle is exploited in soap, to remove grease for the purpose of cleaning. Many different emulsifiers are used in pharmacy to prepare emulsions such as creams and lotions. Common examples include emulsifying wax, cetearyl alcohol, polysorbate 20, and ceteareth 20. Sometimes the inner phase itself can act as an emulsifier, and the result is a nanoemulsion, where the inner state disperses into "nano-size" droplets within the outer phase. A well-known example of this phenomenon, the "Ouzo effect", happens when water is poured into a strong alcoholic anise-based beverage, such as ouzo, pastis, arak, or raki. The anisolic compounds, which are soluble in ethanol, then form nano-size droplets and emulsify within the water. The resulting colour of the drink is opaque and milky white.

Mechanisms of Emulsification

A number of different chemical and physical processes and mechanisms can be involved in the process of emulsification:

- Surface tension theory – according to this theory, emulsification takes place by reduction of interfacial tension between two phases
- Repulsion theory – the emulsifying agent creates a film over one phase that forms globules, which repel each other. This repulsive force causes them to remain suspended in the dispersion medium
- Viscosity modification – emulgents like acacia and tragacanth, which are hydrocolloids, as well as PEG (or polyethylene glycol), glycerine, and other polymers like CMC (carboxymethyl cellulose), all increase the viscosity of the medium, which helps create and maintain the suspension of globules of dispersed phase

Shortening

As compared to Butter, shortening stays more solid at room temperature and because it contains monounsaturated fats and diglycerides, it creams better. The same characteristics help it disperse more evenly in dough, minimize gluten development and efficiently coat flour particles.

Vegetable shortening, unlike butter, can be used in frying as it has a relatively high smoke point. However, you should purchase "pure" shortening, i.e. shortening made entirely from animal or vegetable fats. If "pure" is not on the label, there are additives, and the shortening will have a lower smoke point. Once it is opened, shortening can be stored up for a year in a dark, cool space.

Frequently, shortening compounds are composed of both animal and vegetable fat (derived from various oils) and are mechanically blended to give the final product an acceptable elasticity and satisfactory baking quality. Hydrogenated vegetable oil shortenings, such as Crisco first made in 1911, were developed to withstand higher oven temperatures. Generally, shortening comes in two forms: Solid or Liquid.

Shortening is usually used either in pastry dough or as a butter alternative or complement. But remember that deodorized and neutral tasting shortening does not have the same rich flavour as butter in recipes. Consider adding some butter (30% to 50% of the shortening recommended) to your baked goods for richer flavour. Be careful if substituting for a liquid fat like, vegetable oil, because the recipe can be adversely affected.

The reverse is also true. Swapping in a liquid fat for a solid one can impact the resulting taste and texture.

Because of the milk solids and water in butter that will "set" in your baked product, melted butter won't directly stand in for a liquid fat in a recipe. The best rule of thumb to hold to is to substitute a solid fat for another solid fat, like butter for shortening, and a liquid fat for another liquid fat. Also hydrogenated vegetable oil contains no water. If you want to use it as a substitute for butter, you will need to add in some water as well to achieve a similar result.

Varieties

Solid Shortening: Solid fat, as in butter or shortening, when beaten with sugar can hold air bubbles in its creamy, mass. Solid fats can also be used as a spacer, such as in a pie crust or for leavening, such as in puff pastry. Solid-type shortening is recommended for use in dough because it will not saturate the flour (as the liquid will) and will in turn more thoroughly disperse throughout the dough. Liquid shortening can be used, but only once the dough is more completely formed so the liquid doesn't interfere with the dry flour.

Liquid Shortening: Liquid fats, such as oil, cannot hold air. Liquid-type shortening is suggested for recipes that call for melted shortening, such as some cake and bread recipes.

Culinary Uses

Shortening is a semi-solid, deodorized fat extracted from a vegetable or animal used in food preparations, like Deep Frying, and especially in baking where it gets its name because shortening promotes a "short" or crumbly, flaky texture in crusts or cookies.

Shortening's Smoke Point is 356 to 370°F (180 to 187°C).

To Measure Shortening, use a rubber spatula or spoon and pack it into a dry measuring cup so that the shortening sits level with the graduated line representing the amount you need. For example, for one cup, pack it into the dry measuring cup and level off the cup with a straight edge.

Substitutes

You may also come across Emulsified Shortening which can hold more sugar by weight than traditional shortening holds. The worry about vegetable shortenings is that most on the market are made with partially hydrogenated oils which are high in the bad trans fats and are reputed to be a risk to heart health and healthy metabolizing of other fatty acids. Spectrum Naturals has introduced an organic and trans fat free shortening that you might consider.

Nutrition

Shortening's fat breakdown is 32% Saturated Fat, 11% Monounsaturated Fat, 52% Polyunsaturated Fat.

Control of Quality in Margarine and Shortening Manufacture Haccp

Hazard Analysis and Critical Control Points or HACCP is a systematic preventive approach to food safety from biological, chemical, and physical hazards in production processes that can cause the finished product to be unsafe, and designs measurements to reduce these risks to a safe level. In this manner, HACCP is referred as the prevention of hazards rather than finished product inspection. This standard is a complete food safety and quality management system & the Food Standards Agency also recommends it as the most effective way for food businesses to cover and maintain consumer protection.

Following Processes at Cherubium Lanka are Certified Under HACCP

- Manufacturing of Margarine and Bakery Shortening
- Manufacturing, packing and dispatch of RBD palm olein and coconut oil, physically refined palm olein and coconut oil,

Margarine, shortening (bread fat, frying fat, creaming fat, ice cream fat, palm fat, biscuit dough fat, vegetable shortening, palm oil shortening) , hydrogenated palm oil, hydrogenated palm stearine, compound products (bread compound, bun compound), fat spread & icing sugar

- Packing and dispatch of virgin coconut oil, soyabean oil and sunflower oil

ISO 22000

ISO 22000 specifies requirements regarding the application of food safety management systems in all food chains from producers of food to carriers, retailers and catering establishments.

A HACCP system helps an organisation to identify hazards and put in place control limits at critical points during relevant internal production processes.

HACCP or ISO 22000 certificates awarded from the SLSI implies that chemical, physical and biological hazards encountered during production or processing of the food item, as applicable to the scope of certification controlled to make the food safe for human consumption.

GMP

GMP (Good Manufacturing Practise) is a system to ensure that products meet food safety, quality and legal requirements.

Following areas in the organization is monitored under the standard

1. Establishment - Design, facilities, equipment and maintenance
2. Control of operations
3. Cleaning and sanitation
4. Personal hygiene
5. Transportation
6. Training
7. Product certification
8. Pest control
9. Waste management
10. Packaging and labelling etc

Powdered Fats, Flaked Fats and Fat Powders

Powdered fats are suitable for a variety of food applications and deliver great results every time. They offer a multitude of benefits from improving the taste of a product to helping to overcome difficult production processes. Quick-melting and easily blended, powdered fats can greatly improve product texture.

Powdered fats can be used to help increase the shelf-life of bread. The fine particle size makes it very easy to disperse powdered fats evenly through the dough, even at low addition rates.

When used in fillings and fondant toppings, powdered fats can help keep them glossy, moist and pliable.

Dried foods benefit greatly from the addition of powdered fats. They easily blend with other dried ingredients and when it comes to reconstituting dried foods with milk or water, powdered fats help to emulsify the mix quickly without lumps.

Flaked Fats

Sometimes recipes require the functionality of a hard fat but not the practicality of adding one.

Melting hard fats into a recipe can be a difficult production task. Hard fats are not always easy to handle and can set before fully blended if added into a lower temperature mix. This is when fat flakes can play a vital role in food manufacture. They blend easily with other heat sensitive ingredients before the cooking or baking stage and because of their size they melt out quickly.

In addition to simplifying production techniques, fat flakes can provide multiple benefits across many different products. They can improve textures, add flavour or mimic the structure of other ingredients to create tasty products that cannot be achieved simply with liquid oils.

Fat Powders

The fat powders are made on a basis of dairy and/or vegetable components.

Varieties:

- Fat powder 80% (palm oil or coconut oil on a carrier of glucose in combination with milk proteins)
- Fat powders 50% - 60% (palm oil or coconut oil on a carrier of milk or whey)
- Cream powder 42% - 75% (low or high free fat content)

Applications:

- Bakery mixes
- Ice cream mixes

- Soups and sauces
- Vending

Fat in Biscuit Baking

Biscuit is a term used for a variety of baked, commonly flour-based food products. The term is applied to two distinct products in North America and the Commonwealth of Nations and Europe.

Biscuits are one of those foods that we've somehow convinced ourselves we can't make from scratch, that we need either an insta-pop tin full of doughy chemicals or a southern grandmother at the ready. The very idea of "made from scratch biscuits" conjures thoughts of kneading and a kitchen dusted in flour, something we're all far too busy to attempt on our own, here's a dough boy to the rescue. Nonsense - they are made from a few simple ingredients, very little actual work, and you don't even really need any special equipment.

Fat has several roles in biscuit making. First of all, fat tenderizes. As fat melts with the heat of the oven, it coats gluten strands and starches in the flour and prevents them from becoming too tough. Fats with lower melting points tenderize more than those with higher melting points. Therefore butter is a more effective tenderizer than, say, shortening. Fat is also the key to creating a biscuit's characteristic peelable layers. Solid layers of fat alternating with dough creates flaky layers when the product is baked.

Although many different types of fat make good biscuits, it's important to consider their relative water contents. Fats low in moisture, like shortening, will create more distinct layers, but butter (containing about 15% water) is useful in producing additional steam, which assists in pushing apart the dough layers.

The secret behind great biscuits is fat. That's not really a secret as many bakers know, but how fats are treated in a recipe with leavening, greatly dictates your end result. When fats are used in combination with chemical leaveners, their responsibility is to melt, leaving behind small air pockets. These air pockets are used by the chemical leaveners to release their gases, creating lift in your final product. Knowing how to handle your fats will lead to perfect biscuits every time, no matter what your recipe is!

Examples

Example 1: In the Buttermilk Biscuit recipe, we freeze our butter and grate or pulse it in our food processor so it makes small, hard little balls of fat. These miniature chunks of butter, allow for these small pockets to be created, thus making the interior texture one that has lift, but is full of air pockets when broken open.

Example 2: On the other hand, all the fat comes from the thick cream added to the dough, binding all the dry ingredients together. Because there aren't any "pockets" the gases are released throughout the dough, making the interior still biscuit-y but with holes that are super, super tiny and evenly spread, almost like quick breads such as muffins or banana breads. Their lack of other ingredients doesn't make them taste like that, instead they're silky smooth and waiting to mop up the tail end of the soup left in your dinner bowl.

Example 3: For biscuits that have tender flaky layers that can be peeled apart, piece by piece, the fats are cold lards, shortenings and butters. They are folded and rolled to create lift and layers. Instead of having small pieces or chunks to create holes, by rolling and pressing the dough (like you would make a puff pastry) you create long thin sheets of fat between the rest of your dry ingredients. As they melt away and the leavening happens in their place, flaky layers magically appear.

The Function of Fats in Biscuits

There are different types of fats that can be used for baking. They can be used interchangeably but will lead to a difference of flavour and certain textures. Keep in mind that when you change the type of fat, there may also be a small amount of conversion that has to take place in order for the item to still bake properly.

There are two categories used for baking and cooking as well. These are fats and oils, which also again break down into animal and plant in both of those categories. Animal fat is butter and lard. Plant fat is margarine and shortening.

Oils break down with plant oils such as vegetable and animal and fish oils. Most commonly, baking uses animal and plant fats and sometimes plant oils. Butter serves several purposes. As with sugar, it creates tenderness and aids in a much more moist product. This is why the critical step of creaming the butter and the sugar is key in the beginning of the recipe. Creaming these two items together means you are whipping the items together and creating air. Air then becomes a natural leavening agent.

Butter, margarine or shortening work best for making pie dough, puff pastry dough and many more dough's that have a flaky texture. Ingredients such as oil are not good for this style of recipe since it will not create layers to the dough like the butter does.

Butter or margarine that has a high fat content, about 75 percent or higher, will yield the best final product. Butter is the better choice over margarine or shortening. Spreads that are not true butter do not work well in a baking recipe.

With the fat content being so low and the moisture being so high, the initial step of whipping the butter and sugar together will not have the same results as the butter/sugar combination.

Shortening is another form of a fat used for baking. The name of this solid, shelf stable fat, says it all with its name, used to describe fat's ability to make baked goods tender by impeding the formation of gluten strands. Normally, as bread dough is kneaded, for example, the gluten (wheat protein) begins to join and form long elastic strands, which give strength and a chewy texture to the bread. When fat is added to dough, like in biscuits and pie crusts, the fat gets in the way of the gluten formation, therefore keeping the final product tender and flakey.

Oil leads to mealy, not flaky, pastry and will also lead to a heavier-feeling product. Often times, is the item in baking that most people wish to substitute with applesauce or something lower in fat. Keep in mind, however, that most recipes do require a certain amount of fat just to be successful.

This proves just how critical to baking a lot of these ingredients are and have to be for the optimal results.

Fat Bloom

A great deal has been written about chocolate fat bloom. To date, it is generally accepted that visual fat bloom in chocolate is the cocoa butter that has separated towards the surface. It is caused by:

- Poor (incorrect or incomplete) tempering of the chocolate
- Incorrect cooling methods, including covering cold centres
- The presence of soft fats in the centres of chocolate-covered units
- Warm storage conditions
- The addition to chocolate of fats incompatible with cocoa butter
- Abrasion and finger marking

There is no single universally accepted theory that describes fat bloom in chocolate. Current theories about fat bloom fall into two categories: polymorphic transformation and phase separation theory.

Chapter 3

Water Continuous Emulsions

An emulsion is defined by combining two liquids that will maintain their distinct characteristics after being mixed. When talking about emulsions as applicable in a kitchen, the term emulsion refers to combining fat and water. Culinary emulsion can take two different forms; fat dispersed into water and water dispersed into fat. Common fat in water emulsifications include hollandaise, mayonnaise, aioli, milk, cream, and pan sauces. Water in fat emulsifications are most commonly found in the form of vinaigrettes and whole butter.

In contrast to water and fat maintaining their distinct characteristics when combined, other water friendly liquids like alcohol can never form an emulsion with water because of their ability to freely mix. This is a good thing; if alcohol reacted to water like fat does, then that nice bottle of wine would be nothing more then grape juice with pure alcohol floating on top.

Understanding Emulsions and How They Work

In scientific terms, fat and water are considered "immiscible," or two liquids that don't like to mix. Chemically speaking, when two molecules don't freely mix they will align themselves in such a fashion to touch as little as possible. This alignment is described as "surface tension." It's this surface tension the cook will constantly battle when attempting to make and maintain an emulsification.

In any emulsification there are two phases, the continuous phase and the dispersed phase. The continuous phase can be thought of as the "container" of the emulsification; it is what the other liquid will be "dispersed" into. It's important to keep track of which phase is which because adding more of the continuous phase will thin an emulsion and more of the dispersed phased will thicken an emulsion.

So how do cooks combine fat and water even though they don't like to mix? By using shearing power. Shearing power can be described as a liquid in motion; this can range from shaking, stirring, blending, etc. Shearing power is necessary in the emulsification process to break up the dispersed phase and distribute it throughout the continuous phase. The more shearing power used, the smaller the dispersed phase will become, creating a more stable emulsion. The smaller the dispersed phase, the more of that liquid you can add, yielding a thicker emulsion.

However, this can be taken too far. Like we talked about above, the more dispersed phase contained within an emulsion, the thicker the emulsion will become. But as more of the dispersed phase is added, the dispersed particles become crammed closer together. Eventually there is so much of the dispersed phased that it becomes impossible to keep the individual particles separated. This leads to the dispersed phase first stating to touch, combining, pooling and finally re-creating the natural surface tension. This recreation of surface tension as it applies to cooks is commonly referred to as "broken." For most culinary emulsions, the dispersed phase should not exceed three times the volume of the continuous phase.

Adding too much of the dispersed phase to a given emulsion is only one of many reasons an emulsification will "break." In fact, emulsifications are unstable by nature. Water and oil will always attempt to re-create their natural surface tension, but this inevitability can be prevented by using emulsifiers and stabilizers.

Understanding Emulsifiers and Stabilizers

Emulsifiers and stabilizers are commonly used interchangeably in general kitchen conversations because they both serve the same end goal; stabilizing an emulsion. Emulsifiers however are different from stabilizers and each serves a unique purpose depending on your desired finished product.

There two basic types of emulsifiers a cook can use when creating an emulsion; amino acid chains and phospholipids. Amino acid chains will link together, forming larger chains that we know as proteins. Some of these amino acid chains have both fat friendly and water friendly receptors. A good example are the amino acid chains that make up casein protein, commonly found in egg yolks and milk. Since these amino acid chains by nature are capable of linking fat and water together, by definition they are an emulsifier.

The second form of emulsifiers used in kitchens are phospholipids like lecithin, commonly found in egg yolks and soy. Lecithin is classified as a "surfactant," meaning it has a water friendly head and a fat friendly tail. Lecithin also has the added benefit of have a positively charged tail which makes it an extremely powerful emulsifier, especially when used in a fat-in-water emulsion. The positively charged tail will cause the dispersed fat to actually repel one another, preventing the fat from pooling or "coalescing," which is usually the first step in an emulsion breaking.

Whereas emulsifiers will actually link fat and water together, stabilizers are large molecules that will get in between individual particles contained within the dispersed phase, keeping them from coalescing (pooling together). These "large molecules" can take the form of proteins, starch, pectin, plant particles and food grade gums (a favourite of modern chefs). Because most of these molecules are water soluble, they will usually be utilized with fat-in-water emulsifications.

Inseparable from the concept of large molecules being used as stabilizers is the fact that these large molecules will also "thicken" or add viscosity to an emulsion. In fact, viscosity can help achieve a more stable emulsion especially when starting with a more viscous continuous phase. This is often the case, since food grade gums, starches, plant particles (think purees) will be added to part of the continuous phase before the dispersed phase (commonly fat) is added to the emulsion.

A thicker or more viscous continuous phase will create more drag on the dispersed phase, increasing shearing power, effectively decreasing the particle size of the dispersed phase. As we discussed before, the smaller your particle size, the more stable your emulsion. When using stabilizers you have the added benefit of these large molecules that first aided in forming a more stable emulsion by facilitating a smaller dispersed phase, now getting in the way of the dispersed phase, preventing it from coalescing and braking the emulsion.

How The Emulsification Process Works

In the picture below, you can see the basic emulsification process as it relates to making mayonnaise. This particular emulsification is started by placing acid (commonly lemon juice) and egg yolks in a metal mixing bowl. Shearing power is applied in the form of stirring the contents of the bowl with a whisk.

The egg yolk and lemon juice in this case will be the continuous phase, (the emulsion's "container"), to which we will add our fat (the dispersed phase). It's important in the early stages of the emulsification process to add a very small amount of fat slowly at first, usually just a few drops at a time. The reason this slow start is important is because in the early phases of the emulsification process, the fat has plenty of room to freely move around, avoiding the whisk and instead just floating on top of the water as it naturally prefers to do. Once the fat starts to break up and becomes dispersed throughout the emulsion, additional fat can be added more quickly because the already dispersed droplets will work much like a "mill," breaking the dispersed droplets up into smaller particles, making it much easier to incorporate the dispersed phase in a more continuous stream.

It's important to note however that the more shearing power you are using, the less important this "dispersion mill" becomes. It's much harder for a fat droplet to escape the rapidly churning blade of a high speed blender then that of a hand powered whisk.

To give you a general reference point, the chart below lists some of the most common forms of culinary emulsions while defining both the continuous phase, dispersed phase, emulsifier used, water-emulsifier-fat ratio and shelf stability at normal serving temperatures. All ratios are calculated by weight.

Even though technically the emulsifier is the lecithin and casein contained in the egg yolk, the whole yolk by weight is considered to be the "emulsifier" in this equation to make calculations easier. This means that if you are making a hollandaise sauce and you have 20g lemon juice, you would add to that 20g egg yolks and 100g clarified butter.

Sauce Vierge is a good example of using plant particle and a viscous continuous phase as a stabilizer. In this ratio, the plant particles (tomatoes) are considered the emulsifier for ease of calculation. That means your for every 10g of vinegar or water, you would add 100g plant tissue (tomatoes) and 200g oil.

Milk Fat Globule Structure

More than 95% of the total milk lipid is in the form of a globule ranging in size from 0.1 to 15 um in diametre. These liquid fat droplets are covered by a thin membrane, 8 to 10 nm in thickness, whose properties are completely different from both milkfat and plasma. The native fat globule membrane (FGM) is comprised of apical plasma membrane of the secretory cell which continually envelopes the lipid droplets as they pass into the lumen. The major components of the native FGM, therefore, is protein and phospholipids. The phospholipids are involved in the oxidation of milk. There may be some rearrangement of the membrane after release into the lumen as amphiphilic substances from the plasma adsorb onto the fat globule and parts of the membrane dissolve into either the globule core or the serum. The FGM decreases the lipid-serum interface to very low values, 1 to 2.5 mN/m, preventing the globules from immediate flocculation and coalescence, as well as protecting them from enzymatic action.

It is well known that if raw milk or cream is left to stand, it will separate. Stokes' Law predicts that fat globules will cream due to the differences in densities between the fat and plasma phases of milk. However, in cold raw milk, creaming takes place faster than is predicted from this fact alone. IgM, an immunoglobulin in milk, forms a complex with lipoproteins. This complex, known as cryoglobulinprecipitates onto the fat globules and causes flocculation. This is known as cold agglutination. As fat globules cluster, the speed of rising increases and sweeps up the smaller globules with them. The cream layer forms very rapidly, within 20 to 30 min., in cold milk.

Homogenization of milk prevents this creaming by decreasing the diametre and size distribution of the fat globules, causing the speed of rise to be similar for the majority of globules. As well, homogenization causes the formation of a recombined membrane which is much similar in density to the continuous phase.

Recombined membranes are very different than native FGM. Processing steps such as homogenization, decreases the average diametre of fat globule and significantly increases the surface area. Some of the native FGM will remain adsorbed but there is no longer enough of it to cover all of the newly created surface area. Immediately after disruption of the fat globule, the surface tension raises to a high level of 15 mN/m and amphiphilic molecules in the plasma quickly adsorb to the lipid droplet to lower this value. The adsorbed layers consist mainly of serum proteins and casein micelles.

Fat Destabilization

While homogenization is the principal method for acheiving stabilization of the fat emulsion in milk, fat destabilization is necessary for structure formation in butter, whipping cream and ice cream. Fat destabilization refers to the process of clustering and clumping (partial coalescence) of the fat globules which leads to the development of a continuous internal fat network or matrix structure in the product. Fat destabilization (sometimes "fat agglomeration") is a general term that describes the summation of several different phenomena. These include:

Coalescence: an irreversible increase in the size of fat globules and a loss of identity of the coalescing globules;

Flocculation: a reversible (with minor energy input) agglomeration/clustering of fat globules with no loss of identity of the globules in the floc; the fat globules that flocculate ; they can be easily redispersed if they are held together by weak forces, or they might be harder to redisperse to they share part of their interfacial layers;

Partial Coalescence: an irreversible agglomeration/clustering of fat globules, held together by a combination of fat crystals and liquid fat, and a retention of identity of individual globules as long as the crystal structure is maintained (i.e., temperature dependent, once the crystals melt, the cluster coalesces). They usually come together in a shear field, as in whipping, and it is envisioned that the crystals at the surface of the droplets are responsible for causing colliding globules to stick together, while the liquid fat partially flows between they and acts as the "cement". Partial coalescence dominates structure formation in whipped, aerated dairy emulsions, and it should be emphasized that crystals within the emulsion droplets are responsible for its occurrence.

Dairy Creams

Cream is a dairy product that is composed of the higher-butterfat layer skimmed from the top of milk before homogenization. In un-homogenized milk, the fat, which is less dense, will eventually rise to the top. In the industrial production of cream, this process is accelerated by using centrifuges called "separators". In many countries, cream is sold in several grades depending on the total butterfat content. Cream can be dried to a powder for shipment to distant markets.

Cream skimmed from milk may be called "sweet cream" to distinguish it from whey cream skimmed from whey, a by-product of cheese-making. Whey cream has a lower fat content and tastes more salty, tangy and "cheesy".

Cream produced by cattle (particularly Jersey cattle) grazing on natural pasture often contains some natural carotenoid pigments derived from the plants they eat; this gives the cream a slight yellow tone, hence the name of the yellowish-white colour, cream. Cream from goat's milk, or from cows fed indoors on grain or grain-based pellets, is white.

Types

Different grades of cream are distinguished by their fat content, whether they have been heat-treated, whipped, and so on. In many jurisdictions, there are regulations for each type.

United States

In the United States, cream is usually sold as:

- Half and half (10.5–18% fat)
- Light cream (18–30% fat)
- Light Whipping cream (30–36% fat)
- Heavy cream (36% fat or more)

Not all grades are defined by all jurisdictions, and the exact fat content ranges vary. The above figures are based on the Code of Federal Regulations, Title 21, Part 131

Australia

The Australia New Zealand Food Standards Code - Standard 2.5.2 - Defines cream as milk product comparatively rich in fat, in the form of an emulsion of fat-in-skim milk, which can be obtained by separation from milk. C

ream must contain no less than 350 g/kg of milk fat.

Manufacturers labels may distinguish between different fat contents, a general guideline is as follows:

- Extra light (or 'lite'): 12–12.5% fat.
- Light (or 'lite'): 18–20% fat.
- Pure cream: 35–56% fat, without artificial thickeners.
- Thickened cream: 35–36.5% fat, with added gelatine and/or other thickeners to give the cream a creamier texture, also possibly with stabilizers to aid the consistency of whipped cream (this would be the cream to use for whipped cream, not necessarily for cooking)
- Single cream: Recipes calling for 'single cream' are referring to pure or thickened cream with about 35% fat.
- Double cream: 48–60% fat.

UK

In the United Kingdom, the types of cream are legally defined as followed:

Name	Minimum milk fat	Additional definition	Main uses
Clotted cream	55%	is heat treated	Served as it is. A traditional part of a Cream tea.
Extra-Thick Double cream	48%	is heat treated then quickly cooled	Thickest available fresh cream, spooned onto pies, puddings, and desserts (cannot be poured due to its consistency)
Double cream	48%		Whips easily and thickest for puddings and desserts, can be piped once whipped
Whipping cream	35%		Whips well but lighter, can be piped once whipped
Whipped cream	35%	has been whipped	Decorations on cakes, topping for ice cream, fruit and so on.
Sterilised cream	23%	is sterilised	
Cream or single cream	18%	is not sterilised	Poured over puddings, used in sauces
Sterilised half cream	12%	is sterilised	
Half cream	Uncommon, some cocktails		

Canada

Canadian cream definitions are similar to those used in the United States, except for that of "light cream". In Canada, "light cream" is low fat cream, with 5% or 6% fat. Another form of cream available in Canada is "cereal cream", which is approximately midway between 5% cream and coffee cream in fat content.

Name	Minimum milk fat	Additional definition	Main uses
Whipping cream	35%		Whips well but lighter, can be piped
Table cream	18%		Added to coffee, poured over puddings, used in sauces
Half and half	10-12%		Added to coffee
Light cream	5-10%	Also known as cereal cream	Added to coffee

Japan

In Japan, cream sold in supermarkets is usually between 35% and 48% butterfat.

Switzerland

In Switzerland, the types of cream are legally defined as follows:

English	German	French	Italian	Typical milk fat wt%	Minimum milk fat wt%
Double cream	Doppelrahm	double-crème	doppia panna	45%	45%
Full cream Whipping cream Cream	Vollrahm Schlagrahm Rahm Sahne	crème entière crème à fouetter crème	panna intera panna da montare panna	35%	35%
Half cream	Halbrahm	demi-crème	mezza panna	25%	15%
Coffee cream	Kaffeerahm	crème à café	panna da caffè	15%	15%

Full Cream Whipping Cream Cream

Sour cream and crème fraîche (German: Sauerrahm, Crème fraîche; French: crème acidulée, crème fraîche; Italian: panna acidula, crème fraîche) are defined as cream soured by bacterial cultures. Thick cream (German: verdickter Rahm; French: crème épaissie; Italian: panna addensata) is defined as cream thickened using thickening agents.

Sweden

In Sweden, cream is usually sold as:

- Matlagningsgrädde ("*cooking cream*"), 10-15 %
- Kaffegrädde ("*Coffee cream*"), 10 %
- Vispgrädde (*whipping cream*), 36 - 40 %

Mellangrädde (27 %) is, nowadays, a less common variant. Gräddfil and Creme Fraiche is two common sour cream products.

Processing and Additives

Cream may have thickening agents and stabilisers added. Thickeners include sodium alginate, carrageenan, gelatine, sodium bicarbonate, tetrasodium pyrophosphate, and alginic acid. Other processing may be carried out. For example, cream has a tendency to produce oily globules (called "feathering") when added to coffee. The stability of the cream may be increased by increasing the non-fat solids content, which can be done by partial demineralisation and addition of sodium caseinate, although this is expensive.

Other Cream Products

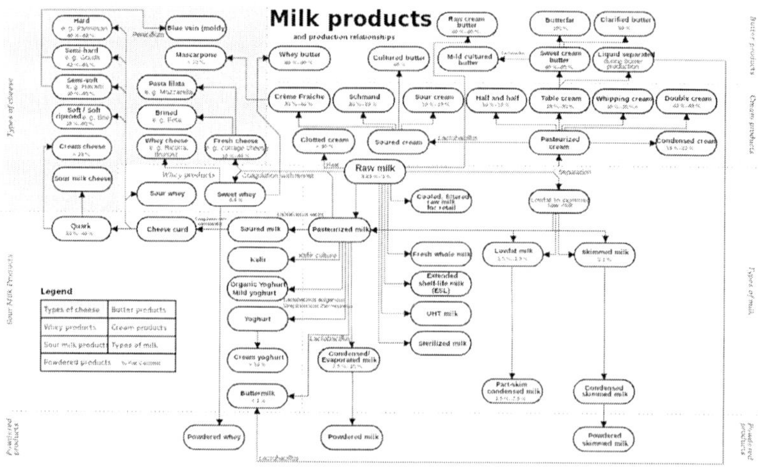

Figure: *Chart of 50 types of milk products and relationships, including cream (click on image to enlarge).*

Butter is made by churning cream to separate the butterfat and buttermilk. This can be done by hand or by machine.

Whipped cream is made by whisking or mixing air into cream with more than 30% fat, to turn the liquid cream into a soft solid. Nitrous oxide may also be used to make whipped cream.

Sour cream, common in many countries including the U.S., Canada and Australia, is cream (12 to 16% or more milk fat) that has been subjected to a bacterial culture that produces lactic acid (0.5%+), which sours and thickens it.

Crème fraîche (28% milk fat) is slightly soured with bacterial culture, but not as sour or as thick as sour cream. Mexican crema (or cream espesa) is similar to crème fraîche.

Smetana is a heavy cream product (15-40% milk fat) Central and Eastern European sweet or sour cream.

Rjome or rømme is Norwegian sour cream containing 35% milk fat, similar to Icelandic sýrður rjσmi.

Clotted cream, common in the United Kingdom, is made through a process that starts by slowly heating whole milk to produce a very high-fat (55%) product. This is similar to Indian malai.

As an Ingredient

Cream is used as an ingredient in many foods, including ice cream, many sauces, soups, stews, puddings, and some custard bases, and is also used for cakes. Irish cream is an alcoholic liqueur which blends cream with whiskey, and often honey, wine, or coffee. Cream is also used in curries such as masala dishes.

Cream (usually light/single cream or half and half) is often added to coffee in the USA. For cooking purposes, both single and double cream can be used in cooking, although the former can separate when heated, usually if there is a high acid content.

Most UK chefs always use double cream or full-fat crème fraîche when cream is added to a hot sauce, to prevent any problem with it separating or "splitting". In sweet and savoury custards such as those found in flan fillings, crème brûlées and crème caramels, both types of cream are called for in different recipes depending on how rich a result is called for.

It is useful to note that double cream can also be thinned down with water to make an approximation of single cream if necessary.

Other Items Called "Cream"

Many non-edible substances are called creams due merely to their consistency: shoe cream is runny, unlike waxy shoe polish; face cream is a cosmetic. There is generally no restriction on describing non-edible products as creams.

Regulations in many jurisdictions restrict the use of the word cream for foods. Words such as creme, kreme, creame, or whipped topping are often used for products which cannot legally be called cream. In some cases foods can be described as cream although they do not contain predominantly milk fats; for example in Britain "ice cream" does not have to be a dairy product (although it must be labelled "contains non-milk fat"), and salad cream is the customary name for a condiment that has been produced since the 1920s and need contain no cream.

Recombined Creams

Recombining some of the common dairy products like milk is relatively easy and straight forward. This is publicly available knowledge. However there are a range of dairy products which are not easily recombined and doing so requires access to specific technology.

Cream of Creams has developed the recombination technology for several dairy products that are now accessible to our dairy plant customers. Below are a set of products which can be added to your product portfolio:

- Whipping cream
- Thick cream
- Sour cream
- Reduced fat cooking cream
- Reduced fat top up cream
- Mascarpone cream cheese
- Ricotta cream cheese

The properties of recombined whipping creams manufactured from anhydrous milk fat, milk solids-non-fat and incorporating various additives have been examined. A blend of 35% butteroil, 7.5% medium heat non-fat milk powder and 0.1% glycerol monostearate is recommended. This formula is relatively insensitive ot homogenizing pressures; two stage homogenization using conditions of 1.4 to 2.1 MPa on the first stage, and 0.7 MPa on the second stage, at 48°C, was found to be satisfactory on the homogenizers used. The properties of the recombined cream produced were not markedly influenced by changes in pH in the range 6.4 - 6.8, or by the non-fat medium-heat milk powder employed.

The formulation is also suitable for manufacture of an ultra-heat-treated recombined whipping cream.

Shelf Life Extension Technology

Besides specific product technology we also offer shelf life extension technology for recombination plants that specifically heat treat their produce by means of pasteurization. Normal pasteurisation will provide around 2 weeks shelf life but the technology to extend this to 6 week's is now accessible to our dairy plant customers.

Ice-cream Mix

Ice cream (derived from earlier iced cream or cream ice) is a frozen food, typically eaten as a snack or dessert, usually made from dairy products, such as milk and cream, and often combined with fruits or other ingredients and flavours. It is typically sweetened with sucrose, corn syrup, cane sugar, beet sugar, and/or other sweeteners. Typically, flavourings and colourings are added in addition to stabilizers. The mixture is stirred to incorporate air spaces and cooled below the freezing point of water to prevent detectable ice crystals from forming. The result is a smooth, semi-solid foam that is solid at very low temperatures (<35 degrees F). It becomes more malleable as its temperature increases.

The meaning of the phrase "ice cream" varies from one country to another. Phrases such as "frozen custard", "frozen yogurt", "sorbet", "gelato" and others are used to distinguish different varieties and styles. In some countries, such as the United States, the phrase "ice cream" applies only to a specific variety, and most governments regulate the commercial use of the various terms according to the relative quantities of the main ingredients.

Products that do not meet the criteria to be called ice cream are labelled "frozen dairy dessert" instead. In other countries, such as Italy and Argentina, one word is used for all variants. Analogues made from dairy alternatives, such as goat's or sheep's milk, or milk substitutes, are available for those who are lactose intolerant, allergic to dairy protein, or vegan.

The most popular flavours of ice cream in North America (based on consumer surveys) are vanilla and chocolate.

Production

Before the development of modern refrigeration, ice cream was a luxury reserved for special occasions. Making it was quite laborious; ice was cut from lakes and ponds during the winter and stored in holes in the ground, or in wood-frame or brick ice houses, insulated by straw. Many farmers and plantation owners, including U.S. Presidents George Washington and Thomas Jefferson, cut and stored ice in the winter for use in the summer. Frederic Tudor of Boston turned ice harvesting and shipping into a big business, cutting ice in New England and shipping it around the world.

Ice cream was made by hand in a large bowl placed inside a tub filled with ice and salt. This was called the pot-freezer method. French confectioners refined the pot-freezer method, making ice cream in a sorbetière (a covered pail with a handle attached to the lid). In the pot-freezer method, the temperature of the ingredients is reduced by the mixture of crushed ice and salt. The salt water is cooled by the ice, and the action of the salt on the ice causes it to (partially) melt, absorbing latent heat and bringing the mixture below the freezing point of pure water. The immersed container can also make better thermal contact with the salty water and ice mixture than it could with ice alone.

The hand-cranked churn, which also uses ice and salt for cooling, replaced the pot-freezer method. The exact origin of the hand-cranked freezer is unknown, but the first U.S. patent for one was #3254 issued to Nancy Johnson on 9 September 1843. The hand-cranked churn produced smoother ice cream than the pot freezer and did it quicker. Many inventors patented improvements on Johnson's design.

In Europe and early America, ice cream was made and sold by small businesses, mostly confectioners and caterers. Jacob Fussell of Baltimore, Maryland was the first to manufacture ice cream on a large scale. Fussell bought fresh dairy products from farmers in York County, Pennsylvania, and sold them in Baltimore. An unstable demand for his dairy products often left him with a surplus of cream, which he made into ice cream. He built his first ice cream factory in Seven Valleys, Pennsylvania, in 1851. Two years later, he moved his factory to Baltimore. Later, he opened factories in several other cities and taught the business to others, who operated their own plants. Mass production reduced the cost of ice cream and added to its popularity.

The development of industrial refrigeration by German engineer Carl von Linde during the 1870s eliminated the need to cut and store natural ice, and, when the continuous-process freezer was perfected in 1926, commercial mass production of ice cream and the birth of the modern ice cream industry was underway.

The most common method for producing ice cream at home is to use an ice cream maker, in modern times, in general, an electrical device that churns the ice cream mixture while cooled inside a household freezer, or using a solution of pre-frozen salt and water, which gradually melts while the ice cream freezes. Some more expensive models have an inbuilt freezing element. A newer method of making home-made ice cream is to add liquid nitrogen to the mixture while stirring it using a spoon or spatula for a few seconds; a similar technique, advocated by Heston Blumenthal as ideal for home cooks, is to add dry ice to the mixture while stirring for a few minutes. Some ice cream recipes call for making a custard, folding in whipped cream, and immediately freezing the mixture.

Ingredients and Standard Quality Definitions

Figure: *Black sesame soft ice cream, Japan*

In the USA, ice cream may have the following composition:

- greater than 10% milkfat and usually between 10% and as high as 16% fat in some premium ice creams
- 9 to 12% milk solids-not-fat: this component, also known as the serum solids, contains the proteins (caseins and whey proteins) and carbohydrates (lactose) found in milk
- 12 to 16% sweeteners: usually a combination of sucrose and glucose-based corn syrup sweeteners
- 0.2 to 0.5% stabilisers and emulsifiers
- 55% to 64% water, which comes from the milk or other ingredients.

These compositions are percentage by weight. Since ice cream can contain as much as half air by volume, these numbers may be reduced by as much as half if cited by volume. In terms of dietary considerations, the percentages by weight are more relevant. Even the low-fat products have high caloric content: Ben and Jerry's No-Fat Vanilla Fudge contains 150 calories (630 kJ) per half-cup due to its high sugar content.

Commercial Ice Cream Mixes

Ever wondered how small ice cream shops and restaurants make their homemade ice cream? Well, it turns out many of them don't actually beat together the eggs and sugar, and then heat the milk and cream, and go through the whole process like we do at home.

They just pour a pre-made mix into the ice cream machine and let it churn away. And they get that ice cream mix from a local dairy.

The mix typically has cream, milk, and sugar, but without the flavouring, which can be added depending on what they're making.

- 14% Mix – This was the official ice cream mix for an ice cream company with locations around the world. If you live in the U.S., you've almost certainly had it. This particular one was unflavoured, or in other words, just a Sweet Cream mix.
- 10% Mix – Since ice cream in the U.S. has to have at least 10% butterfat to be called "ice cream", this would be considered a "low fat" or "reduced fat" ice cream. It was also unflavoured, and probably used by lots of restaurants, schools, etc.
- 5% Mix – Another official mix for a very famous international company, and this one's been around a very long time. This mix has a very distinct, vanilla-like flavour that you can almost immediately identify when you first taste it. Since it's only 5% butterfat, it's considered a soft serve mix, which would normally be eaten right out of the machine.
- 3.5% Vanilla Mix – This would be considered an "ice milk", since it has the same butterfat content as whole milk. This would also be used in a soft serve machine, and eaten fresh. Think about soft serve machines you've seen at restaurants giving away free cones, etc.
- 3.5% Chocolate Mix - Same as the vanilla, but with the chocolate flavour already mixed in to make things easier for the company using it.

  The vanilla and chocolate mixes might even be used in the same machine for swirls.

Manufacturers Produce ice Cream to Meet the Requirements of Consumers.

In many situations the ice cream manufacturer will have a final product specification to meet. An ice cream recipe must be produced from this specification. One element of this specification will be the chemical composition of the final product including any legal compositional requirements e.g. for fat and milk protein. In the UK ice cream must contain a minimum of 5% fat and a minimum of 2.5% milk protein. A compositional specification will typically specify the fat, milk solids not fat (MSNF), sugar, emulsifier and stabiliser concentrations in the final product. In this situation the manufacturer will select ingredients that can supply the above components, blend these and then process to produce a finished ice cream. Calculations will be required to ensure that the ingredients are correctly formulated to meet the final product specification.

In other situations the manufacturer may want to produce a new or improved product and will have to devise their own product compositional specification. This specification must be developed so that the fat, sugar and MSNF components are balanced.

Fat /Sugar Balance

Fat adds certain taste and other qualities to food. In ice cream fat is balanced by sugar and there is a relationship between fat concentration and sugar concentration.

The relationship between fat and sugar concentrations is also influenced by the type of freezer used.

Table : Recommended fat and Sugar Concentrations for Ice Cream Mixes Using Various Types of Freezer

Freezer Type	*Fat concentration (% w/w)*	*Sugar concentration (% w/w)*
Vertical	6	12
Vertical	7	12.5
Vertical	8	13
Horizontal	9	13.5
Horizontal	10	14
Continuous	10	14
Continuous	12	15

Note the fat/sugar ratios above should be regarded as approximate and will need to be modified to meet differing consumer requirements.

Also, sweetners other than sucrose have become more extensively used. However, from a product development perspective the information.

Guinard *et al.* studying the influence of fat and sugar in a vanilla ice cream used university students to taste and rate on a nine-point hedonic scale the texture and mouth feel, flavour (taste and odour), and overall degree of liking for nine samples of vanilla ice cream varying in sugar (8.94 to 18.81%, w/w) and fat (8.73 to 19.30%, w/w) concentration.

The hedonic ratings differed significantly among samples, and the best-liked sample for texture and mouth feel, flavour, and overall degree of liking contained 13.54 % sugar and 14.99 % fat. Response surface methodology, in simple terms a three dimensional graphical analysis of overall degree of liking versus sugar and fat concentrations, was used to relate hedonic ratings to sugar and fat percentages in the ice cream.

Dome-shaped response surfaces, were obtained for all three degree of liking parametres, and optimal sugar and fat, respectively, were 13.16 % and 14.02 % for degree of liking of texture and mouth feel, 14.07 % and 15.35 % for degree of liking of flavour, and 14.30 % and 14.77 % for overall degree of liking.

Preparation of Dressings

Sauces for salads are often called "dressings". The concept of salad dressing varies across cultures.

In Western culture, there are two basic types of salad dressing:

- Vinaigrette;
- Creamy dressings, usually based on mayonnaise or fermented milk products, such as yogurt, sour cream (crème fraîche, smetana), buttermilk;

Vinaigrette is a mixture (emulsion) of salad oil and vinegar, often flavoured with herbs, spices, salt, pepper, sugar, and other ingredients. It is also used as a sauce or marinade.

In North America, mayonnaise-based Ranch dressing is most popular, with vinaigrettes and Caesar-style dressing following close behind. Traditional dressings in France are vinaigrettes, typically mustard-based, while sour cream (smetana) and mayonnaise are predominant in eastern European countries and Russia. In Denmark, dressings are often based on crème fraîche. In southern Europe, salad is generally dressed by the diner with oil and vinegar.

In Asia, it is common to add sesame oil, fish sauce, citrus juice, or soy sauce to salad dressings.

The following are examples of common salad dressings:

- Blue cheese dressing
- Caesar dressing
- Extra virgin olive oil
- French dressing
- Ginger dressing
- Honey Dijon
- Hummus
- Italian dressing
- Louis dressing
- Ranch dressing
- Russian dressing
- Tahini
- Thousand Island dressing
- Vinaigrette
- Wafu dressing

Toppings and Garnishes

Popular salad garnishes are nuts, croutons, anchovies, bacon bits (real or imitation), garden beet, bell peppers, shredded carrots, diced celery, watercress, sliced cucumber, parsley, sliced mushrooms, sliced red onion, radish, french fries, sunflower seeds (shelled), real or artificial crab meat (surimi) and cherry tomatoes. Various cheeses, berries, seeds and other ingredients can also be added to green salads. Cheeses, in the form of cubes, crumbles, or grated, are often used, including blue cheese, Parmesan cheese, and feta cheese. Colour considerations are sometimes addressed by using edible flowers, red radishes, carrots, various colours of peppers, and other colourful ingredients.

Mayonnaise

Mayonnaise, often abbreviated as mayo, is a thick, creamy sauce often used as a condiment. It is a stable emulsion of oil, egg yolks and either vinegar or lemon juice, with many options for embellishment with other herbs and spices. Lecithin in the egg yolk is the emulsifier.

Mayonnaise varies in colour, but is often white, cream, or pale yellow. It may range in texture from that of light cream to a thick gel. In countries influenced by French culture, mustard is also a common ingredient, but the addition of mustard turns the sauce into a remoulade with a different flavour and the mustard acts as an additional emulsifier. In Spain, Portugal, Italy and Greece, olive oil is used as the oil and mustard is never included.

Commercial egg-free mayonnaise-like spreads are available for people who want to avoid animal fat and cholesterol, or who are allergic to eggs.

Figure: *Fresh homemade mayonnaise*

Preparation

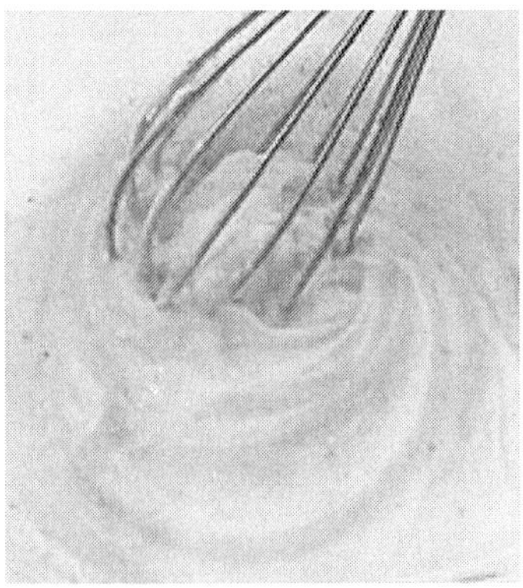

Figure: *Making mayonnaise with a whisk.*

Mayonnaise can be made by hand with a mortar and pestle, whisk or fork, or with the aid of an electric mixer or blender. Mayonnaise is made by slowly adding oil to an egg yolk, while whisking vigorously to disperse the oil. The oil and the water in yolks form a base of the emulsion, while lecithin and protein from the yolks are the emulsifiers that stabilize it. Additionally, a bit of a mustard may also be added to sharpen its taste, and further stabilize the emulsion. Mustard contains small amounts of lecithin. If vinegar is added directly to the yolk it can emulsify more oil, thus making more mayonnaise.

For large-scale preparation of mayonnaise where mixing equipment is being employed the process typically begins with the dispersal of eggs, either powdered or liquid, into water. Once emulsified, the remaining ingredients are then added and vigorously mixed until completely hydrated and evenly dispersed. Oil is then added as rapidly as it can be absorbed. Though only a small part of the total, ingredients other than the oil are critical to proper formulation. These must be totally hydrated and dispersed within a small liquid volume, which can cause difficulties including emulsion breakdown during the oil-adding phase. Often a long agitation process is required to achieve proper dispersal/emulsification, presenting one of the trickiest phases of the production process. Though, as technology in the food industry advances, processing has been shortened drastically allowing roughly 1000 litres to be produced in 10 minutes.

Composition

Homemade mayonnaise can approach 85% fat before the emulsion breaks down; commercial mayonnaise is more typically 70% to 80% fat. "Low fat" mayonnaise products contain starches, cellulose gel, or other thickeners to simulate the texture of real mayonnaise.

Commercial producers either pasteurize the yolks, freeze them and substitute water for most of their liquid, or use other emulsifiers. They also typically use soybean or rapeseed oil, for its lower cost, instead of olive oil. Some recipes, both commercial and homemade, use the whole egg, including the white.

Uses

Figure: *Mayonnaise from the Zaan district, North-Holland, Netherlands and French fries.*

Chile

Chile is the world's third major per capita consumer of mayonnaise and first in Latin America. Since mayonnaise became widely accessible in the 1980s Chileans have used it on locos, completos, French fries, and on boiled chopped potatoes, a salad commonly known as "papas mayo".

Europe

In European countries, especially Belgium and the Netherlands, mayonnaise is often served with *pommes frites*, French fries, or chips. It is also served with cold chicken or hard-boiled eggs in France, Poland, the UK, Benelux, Hungary, Austria, the Baltic States and Eastern Europe.

Guidelines issued in September 1991 by Europe's Federation of the Condiment Sauce Industries recommend that oil and liquid egg yolk levels in mayonnaise should be at least 70% and 5% respectively. The Netherlands incorporated this guideline in 1998 into the law "Warenwetbesluit Gereserveerde aanduidingen" in article 4. Most available brands easily exceed this target.

North America

Commercial mayonnaise sold in jars originated in Philadelphia in 1907 when Amelia Schlorer decided to start selling her own mayonnaise recipe originally used in salads sold in the family grocery store. Mrs. Schlorer's Mayonnaise was an instant success with local customers and eventually grew into the Schlorer Delicatessen Company. Around the same time in New York City, a family from Vetschau, Germany, at Richard Hellmann's delicatessen on Columbus Avenue, featured his wife's homemade recipe in salads sold in their delicatessen. The condiment quickly became so popular that Hellmann began selling it in "wooden boats" that were used for weighing butter. In 1912, Mrs. Hellmann's mayonnaise was mass-marketed and later was trademarked in 1926 as Hellmann's Blue Ribbon Mayonnaise.

At about the same time that Mrs. Schlorer's and Hellmann's Mayonnaise were thriving on the East Coast of the United States, a California company, Best Foods, introduced their own mayonnaise, which turned out to be very popular in the western United States. In 1932, Best Foods bought the Hellmann's brand. By then, both mayonnaises had such commanding market shares in their own half of the country that it was decided that both brands be preserved. The company is now owned by Unilever.

In the southeastern part of the United States, Mrs. Eugenia Duke of Greenville, South Carolina, founded the Duke Sandwich Company in 1917 to sell sandwiches to soldiers training at nearby Fort Sevier. Her homemade mayonnaise became so popular that her company began to focus exclusively on producing and selling the mayonnaise, eventually selling out to the C. F. Sauer Company of Richmond, Virginia, in 1929. Duke's Mayonnaise remains a popular brand of mayonnaise in the Southeast, although it is not generally available in other markets.

In addition to an almost ubiquitous presence in American sandwiches, mayonnaise forms the basis of northern Alabama's signature white barbecue sauce. It is also used to add stability to American-style buttercream and occasionally in cakes as well.

Japan

Japanese mayonnaise is typically made with apple cider vinegar or rice vinegar and a small amount of MSG, which gives it a different flavour from mayonnaise made from distilled vinegar. It is most often sold in soft plastic squeeze bottles. Its texture is thicker than most Western commercial mayonnaise. A variety containing karashi (Japanese mustard) is also common.

Apart from salads, it is popular with dishes such as okonomiyaki, takoyaki and yakisoba and may also accompany katsu and karaage. It is sometimes served with cooked vegetables, dabbed on sushi or mixed with soy sauce, hot/spicy chili oil or wasabi and used as dips. In the Tôkai region, it is a frequent condiment on *hiyashi chûka* (cold noodle salad). Many fried seafood dishes are served with a side of mayonnaise for dipping. It is also common in Japan to use mayonnaise on pizza. Mayonnaise is also often used for cooking where it can replace butter or oil when frying vegetables or meat.

Kewpie (Q.P.) is the most popular brand of Japanese mayonnaise advertised with a Kewpie doll logo. It is made with egg yolks instead of whole eggs, and the vinegar is a proprietary blend containing apple and malt vinegars.

Russia

Mayonnaise is very popular in Russia where it is made with sunflower seed oil which gives it a very distinctive flavour. A 2004 study showed that Russia is the only market in Europe where mayonnaise is sold more than ketchup by volume. It is used as a sauce in the most popular salads in Russia, such as Russian salad, or Olivier salad, and dressed herring and also many others. Leading brands are Calve (marketed by Unilever) and Sloboda (marketed by Efko).

Furthermore, in many eastern European countries (Russia, Ukraine, Belarus, etc.), one can find different commercial flavours of mayonnaise, such as olive, quail-egg, and lemon.

China

There are many terms for mayonnaise in China. Imported brands of mayonnaise and whipped dressing can often be found at multicultural supermarkets in China such as Carrefour.

One reason that mayonnaise is called salad oil is because it is commonly found in potato salad and Olivier salad, which may have become popularized in China through Soviet cuisine.

As a Base for Other Sauces

Mayonnaise is the base for many other chilled sauces and salad dressings. For example:

- Fry sauce is a mixture of mayonnaise, ketchup or another red sauce (e.g. Tabasco sauce, Buffalo wing sauce, or one of many smoky barbecue sauces popular in the Northwestern United States), spices, and sometimes a strong tasting salty liquid (such as Worcestershire or soy sauce) is added to balance out the sweeter red sauces. Commonly eaten on French fries in Utah, Idaho, eastern Washington and rural Oregon.
- Marie Rose sauce combines mayonnaise with tomato sauce or ketchup, cream, flavourings and brandy. In North America, a processed version of Marie-Rose, called "Russian dressing" sometimes uses mayonnaise as a base. However, most homemade varieties and nearly all commercial brands of "Russian dressing" use little or no mayonnaise as a base. They are very dark red and sweet dressings made with vegetable oil, tomato paste, vinegar, sugar, and a variety of herbs and spices (often including mustard).
- Ranch dressing is made of buttermilk or sour cream, and minced green onions, along with other seasonings, and is sometimes mixed with mayonnaise although that is not a traditional ingredient.
- Rouille is aïoli with added saffron, red pepper or paprika.
- Salsa golf created in Argentina is mayonnaise with ketchup as well as spices such as red pepper or oregano.
- Sauce rémoulade, in classic French cuisine is mayonnaise to which has been added mustard, gherkins, capers, parsley, chervil, tarragon, and possibly anchovy essence. An industrially made variety is popular in Denmark and Sweden with French fries and fried fish. It is quite different from most of the remoulade sauces that are frequently found in Louisiana and generally do not have a mayonnaise base.

- Tartar sauce is mayonnaise spiced with pickled cucumbers and onion. Capers, olives, and crushed hardboiled eggs are sometimes included. A simpler recipe calls for only pickle relish to be added to the mayonnaise.
- Thousand Island dressing is a salmon-pink dressing that combines tomato sauce and/or tomato ketchup or ketchup-based chili sauce, minced sweet pickles or sweet pickle relish, assorted herbs and spices (usually including mustard), and sometimes including chopped hard-boiled egg—all thoroughly blended into a mayonnaise base.
- Certain variations of honey mustard are based on mayonnaise and are made by combining mayonnaise with plain mustard, brown sugar, and lemon juice.

Nutritional Aspects

Commercially made mayonnaise may contain sucrose, high fructose corn syrup, citric acid, thickeners, emulsifiers, EDTA, flavour enhancers, and water. Such mixtures allow for the production of products that are low in fats and/or sugars. Commercial mayonnaise is also readily available without these additional ingredients.

A typical formulation for commercially made mayonnaise (not low fat) can contain as much as 80% vegetable oil, usually soybean but sometimes olive oil. Water makes up about 7% to 8% and egg yolks about six percent. Some formulas use whole eggs instead of just yolks. The remaining ingredients include vinegar (4%), salt (1%) and sugar (1%). Low-fat formulas will typically decrease oil content to just 50% and increase water content to about 35%. Egg content is reduced to 4% and vinegar to 3%. Sugar is increased to 1.5% and salt lowered to 0.7%. Gums or thickeners (4%) are added to increase viscosity, improve texture and ensure a stable emulsion.

Figure: *Vegetarian taro veggie burger with relish, tomato, salad and Vegenaise.*

There are several ways to prepare mayonnaise, but on average mayonnaise is approximately 700 kilocalories (2,900 kJ) per 100 grams of product. This makes mayonnaise a calorically dense food.

- There are egg-free mayonnaise-like spreads available for people who want to avoid animal fat and cholesterol, or who have egg allergies. These are also suitable for vegans, and for religious vegetarians who abstain from egg consumption, such as followers of Hindu vegetarianism. Well-known brands include Nayonaise and Vegenaise in North America, and Plamil Egg Free in the UK.

Defects in Ice Cream

Defects may appear in ice cream due to following reasons:

- Use of low quality ingredients
- Improper mixing of ingredients
- Faulty method of manufacture
- Faulty method of storage and transport

Quality of ice cream is judged for its delicate and pleasing flavour, smooth body and texture and attractive colour. The reasons for the development of defects in ice cream and the preventive actions to be taken to remove the defects are discussed in the following paragraphs.

- Colour and appearance Defects and Preventive Measures:
- Un-natural colour in ice cream develop due to following reasons:
- Careless-ness in adding colour
- Improper use of colour
- Use of foreign materials
- Uneven colour results if the colour is not added and mixed properly.
- Excessive colour is due to addition of too much colour.

To avoid any of the above colour defects in ice cream correct amount of colour should be calculated, added and mixed thoroughly. Also the added colour should reflect the flavour characteristic of the ice cream. Cleanliness and neatness of the pakages should also be carefully observed.

- Melting quality defects and Preventive Measures
- Curdy melt down or curdled melt down: It indicates high acidity in the ice cream mix or the effect of any other factor which causes instability of milk proteins. This defect makes

the consumer feel that inferior quality materials were used in the preparation of ice cream. The defect may be corrected by:

- Using fresh dairy products
- Avoiding use of any product that might disturb natural salt balance of the mix
- Avoiding freezing of high acid mixes
- Ice cream does not melt: Ice cream shows resistance to melting, retains shape when warmed and takes longer time to melt. The cause and remedial measures of the defect are discussed under soggy body defecs.
- Slow meltdown: It indicates use of excessive stabilizer (over stabilization) or faulty processing of the mix. This condition may be corrected by taking the following measures:
- Reducing amount of stabilizer
- Using fresh dairy products
- Homogenizing the mix at proper temperature and pressure
- Whey leakage: When ice cream melts separation of whey occurs. This defect may occur due to poor quality of ice cream mix or improper balancing of mix. This defect can be corrected by:
- Using good quality dairy products
- Balancing the constituents carefully
- Using better quality stabilizer
- Foamy meltdown: This is caused by incorporation of too much air in the ice cream during freezing (excessive over run). Hence the defect may be corrected by reducing the over run and reducing the amount of emulsifier or egg products.
- Body and texture defects and Preventive Measures

The body of ice cream refers to firmness and consistency while texture refers to the fineness of the product, which is dependent upon the number, size, shape and arrangement of ice crystals and other particles. The ideal body of ice cream is that which is produced by the correct proportion of milk solids together with proper over run and which melts fairly rapidly at room temperature to a smooth liquid similar in consistency and appearance to sweet cream containing about 40 % fat.

- Crumbly body: It is a condition in which ice cream does not hold together properly. It lacks cohesion or breaks apart very easily. It is associated with low total solids content, in-sufficient stabilization, excessive over run, imperfect homogenization and large air cells. This defect can be remedied by:

- Increasing the total solids content
- Increasing the stabilizer
- Decreasing the over run
- Soggy body: Ice cream with a dense body and wet appearance is an indication of soggy body. It is due to a low over run, high concentration of sugar that lowers the freezing point, excessive use of stabilizer or delayed packaging of ice cream after freezing. This defect is also similar to other defects such as gummy, doughy, sticky, pasty and gluey. Excessive stabilization or high total solids content produce a chewy or gummy body while certain types of gums and syrups cause pasty or sticky body. Soggy body and other related defects contribute to high melting resistance. These defects can be corrected by:
- Maintaining correct over run
- Reducing the quantity of stabilizer
- Use of a combination of stabilizers
- Reducing the sugar content
- Weak body: Ice cream lacks firmness or chewiness and is invariably accompanied by rapid melting. This defect is due to low total solids content combined with insufficient stabilization. This defect can be corrected by:
- Increasing the total solids content
- Adding correct amount of stabilizer
- Properly ageing the mix
- Texture of ice cream: Ice cream having an ideal texture will be very smooth,i.e., the solid particles too small to detect in mouth.
- Buttery texture: This defect appears in ice cream when lumps of butter- fat are detected easily in the mouth. This defect is due to high fat content, inadequate homogenization, mix entering the freezer at a higher temperature and partial churning of cream due to slow freezing. This defect may be corrected by:
- Homogenizing the mix properly
- Ageing the mix properly

- Freezing the mix quickly
- Coarse or icy texture: When the ice crystals are large or not uniform in size or that the air cells are too large in ice cream, icy texture is indicated. This defect is most commonly noticed in retailer's shop. Large ice crystal formation in ice cream may be due to
- Insufficient stabilizer
- Slow freezing in the freezer
- Slow freezing in the hardening room
- Insufficient ageing of mix

Formation of Large air Cells May be Due to

- Type of freezer used
- Use of mix of low whipping ability

Sometimes small ice particles are found in ice cream. These ice particles are formed when water droplets get into ice cream. This defect may be eliminated by

- Increasing total solids in the ice cream
- Increasing or adding correct amount of stabilizer
- Freezing quickly and drawing ice cream at a lower temperature from freezer
- Hardening ice cream quickly
- Reducing storage period
- Avoiding frequent heat shocks to ice cream
- Fluffy texture: This defect is due to presence of large amount of air cells. It is due to incorporation of an excessive amount of air either as large or small air cells. This defect may be expected when the total solids content in ice cream is not more than one third of over run. When the air cells are large and the amount of air is excessive the texture is sometimes described as snowy or flaky. This defect may be removed by
- Decreasing over run
- Increasing total solids
- Decreasing the amount of emulsifiers
- Sandy texture: This defect is easily detected by the presence of rough sand like particles in mouth when ice cream is tasted. This defect may be due to presence of more non-fat milk solids, which contribute more lactose. Higher amount of sugar may also lead to lactose crystallization. This defect can be eliminated by

- Reducing milk solids not fat content
- Replacing part of cane sugar with other sweetening agents like dextrose, corn syrup etc
- Maintaining uniformly low temperature during hardening and storage.
- Flavour Defects and Preventive Measures:

Ice cream flavour is essentially a blend of flavours from different ingredients used in its preparation. The dairy products and flavouring materials used have profound effect on the delicate flavour characteristics of ice cream. Any defect in these materials will therefore influence significantly the flavour of ice cream.

Defects Due to Flavouring Materials

- High flavour: Excess of flavouring material will impart a sharp bitter flavour in ice cream. Poor quality flavouring material also imparts bitter flavour. This can be corrected by checking the quality of flavouring materials and adding correct amount to get the desired delicate flavour.
- Low flavour: It is due to insufficient amount of flavouring material. It requires addition of correct amount of flavouring material to overcome the defect.
- Harsh flavour: It is due to use of inferior flavouring substances. These substances or artificial extracts lack fine flavour . Also use of excess flavour cause this defect. Therefore correct quantity of superior flavours should be used.
- Acid flavour: This is caused by the presence of excessive amount of lactic acid. This can be corrected by:

Using Fresh Dairy Products

Prompt and Efficient Cooling of Mix: Avoiding prolonged storage of the mix at high storage temperature.

- Bitter flavour: This defect may be caused by the use of inferior quality ingredients. The defect may be controlled by:

Using good / fresh Extracts: Avoiding use of dairy products stored for long time at low temperature as certain type of bacteria produce bitter flavour under these conditions.

- ***Cooked Flavour:*** This is caused by over heating the mix or using over heated concentrated dairy products. The defect can be prevented by: Carefully controlling pasteurization temperature Using concentrated dairy products without cooked flavour

- ***Flat Flavour:*** This results from the use of insufficient flavours, sugars or milksolids. By adding required quantities of various ingredients this defect can be corrected.
- ***Un-natural Flavour:*** This is indicated by the flavour not typical of ice cream.The defect can be controlled by using good quality dairy products and flavouring materials.Synthetic flavours that are not correct imitation of true flavours are also considered un-natural. Similarly flavours due to poor gelation, fermented syrups,over-ripe or under-ripe fruits and rancid nuts are un-natural. Therefore use of such ingredients should be avoided to control this defect.
- ***Oxidized Flavour:*** This defect is also known as tallowy or cardboard flavour and results from use of old or stored dairy ingredients especially fat rich products.This defect can be remedied by use of fresh dairy products and pasteurizing the mix at slightly higher temperature than recommended.
- ***Shrinkage:*** Besides flavour and body texture defects one another defect is also noticed in ice cream, especially in retail shops. Sometimes when well-filled packages are hardened and stored, the volume of ice cream shrinks leaving space either at the top or on the sides of the package. Cups then do not appear full. This defect is known as Shrinkage defect. The main reason for this defect to occur appears to be excessive over run in ice cream and too much fluctuation in storage temperature. Maintaining constant temperature during storage and avoiding rough handling during transportation will prevent this defect to a greater extent in ice cream.

Chapter 4

Oil Modification Processes

Processing can remove the components of edible oils which may have negative effects on taste, stability, appearance or nutritional value. To the extent possible, processing should preserve tocopherols and prevent chemical changes in the triacyglycerols.

Rural oil extraction usually occurs near the areas of raw material production. This provides the smallscale processor with access to raw materials, helps to ensure that perishable oil crops are processed quickly, and reduces transport costs. For rural communities and the urban poor, unrefined vegetable oils contribute significantly to the total amount of oil consumed. Crude oils are affordable to low-income groups and serve as important sources of b -carotene and tocopherols.

To maintain the quality of the raw material, care is needed during and after the harvesting of oilbearing fruits that are perishable and susceptible to fat breakdown. Bruising of fresh palm fruits accelerates lipase activity leading to fat degradation. Oil-bearing crops such as sheanuts are prone to mould infestation during storage. This is curtailed by heat treatment: steaming or boiling, coupled with sun-drying to reduce the moisture content.

Storage. The moisture content of oil seeds and nuts influences the quality of raw materials over time. In most rural operations, sun-drying reduces the moisture content of oil seeds to below 10 percent. Adequate ventilation or aeration of the seeds or nuts during storage ensures that low moisture levels are maintained and microbial development is avoided. This is important in the storage of groundnuts which are highly susceptible to aflatoxin contamination through the growth of Aspergillusflavus. Since aflatoxins and pesticides are not removed by rural extraction techniques, microbial contamination and the application of insecticides should be avoided. There is a need for storage practices which are affordable and available to the small-scale processor. Perishable raw materials such as palm fruits should be processed as soon as possible after harvesting.

In humid developing countries, the sun-drying of oil seeds with a high moisture content, such as mature coconut, is slow and inefficient. Such conditions promote mould growth which results in high free fatty acid levels and poor organoleptic qualities. Coconut oil for human consumption should be obtained soon after harvest.

Pre-treatment. The first operation after harvesting involves sterilization and heat treatment by steaming or boiling, this inactivates lipolytic enzymes which could cause rapid degradation of the oil and facilitates the pulping of the mesocarp for oil extraction. "Sterilised" palm fruits are pulped in a wooden pestle and mortar or mechanised digestor.

Decortication or shelling separates the oil-bearing portion of the raw material and eliminates the parts that have little or no nutritional value. Small-scale mechanical shellers are available for kernels and nuts although manual cracking is still prevalent.

Most oil seeds and nuts are heat-treated by roasting to liquify the oil in the plant cells and facilitate its release during extraction. All oil seeds and nuts undergo this treatment except palm fruits for which "sterilization" replaces this operation.

To increase the surface area and maximize oil yield, the oil-bearing part of groundnuts, sunflower, sesame, coconut, palm kernel and sheanuts is reduced in size. Mechanical discattrition mills are commonly used in rural operations.

Extraction. In oil extraction, milled seed is mixed with hot water and boiled to allow the oil to float and be skimmed off. The milled oil seed is mixed with hot water to make a paste for kneading by hand or machine until the oil separates as an emulsion. In groundnut oil extraction, salt is usually added to coagulate the protein and enhance oil separation.

A large rotating pestle in a fixed mortar system can be powered by motor, humans or animals to apply friction and pressure to the oil seeds to release oil at the base of the mortar. Other traditional systems used in rural oil extraction include the use of heavy stones, wedges, levers and twisted ropes. For pressing, a plate or piston is manually forced into a perforated cylinder containing the milled or pulped oil mass by means of a worm. The oil is collected below the perforated chamber. A variety of mechanical expellers have been designed. The pre-heated raw material is fed into a horizontal cylinder by a wormshaft. By means of an adjustable choke, internal pressure which is built up in the cylinder ruptures the oil cells to release the oil.

Dehydration. By boiling in shallow pans, traces of water in crude oil are removed after settling. This is common in all rural techniques which recognize the catalytic role of water in the development of rancidity and poor organoleptic qualities.

Pressed cakes. A by-product of processing, the pressed cake, may be useful depending on the oil extraction technique applied. Cakes from water-extracted oil are usually depleted of nutrients. Other traditional techniques, for instance, those used for groundnut and copra ensure that the by-products, if handled with care, are suitable for human consumption.

Traditional technologies. In many countries, traditional processes for producing oil are very important, especially among communities which have easy access to raw oleaginous materials. Traditional processing tends to be environmentally sound and the skills required are family or group activities, involving women in particular. In a changing industrial atmosphere, these positive features have been outweighed by the negative aspects of traditional processing such as small production capacities, poor economies of scale, high expenditures of energy and time, and the cost of transporting oils to markets.

Large-scale Production

Storage. Many steps in industrial processing find their origin in the traditional processes. In large-scale operations, oilseeds are dried to less than 10 percent moisture. They may be stored for prolonged time periods under suitable conditions of aeration with precautions against insect and rodent infestation. Such storage reduces mould infection and mycotoxin contamination and minimizes biological degradative processes which lead to the development of free fatty acids and colour in the oil.

Oil-bearing fruits such as olive and palm are treated as quickly as possible. Palm is sterilized as a first step in processing. Adipose tissues and fish-based raw materials (that is, the body or liver) are rendered within a few hours by boiling to destroy enzymes and prevent oil deterioration.

Processing. Oilseeds are generally cleaned of foreign matter before dehulling. The kernels are ground to reduce size and cooked with steam, and the oil is extracted in a screw or hydraulic press. The pressed cake is flaked for later extraction of residual fat with solvents such as "food grade" hexane. Oil can be directly extracted with solvent from products which are low in oil content, that is, soybean, ricebran and corn germ.

After sterilization, oil-bearing fruits are pulped (digested) before mechanical pressing often in a screw press. Palm kernels are removed from pressed cakes and further processed for oil. Animal tissues are reduced in size before rendering by wet or dry processes. After autoclaving, tissues of fish are pressed and the oil/water suspension is passed through centrifuges to separate the oil.

Oil Refining. Refining produces an edible oil with characteristics that consumers desire such as bland flavour and odour, clear appearance, light colour, stability to oxidation and suitability for frying. Two main refining routes are alkaline refining and physical refining (steam stripping, distillative neutralisation) which are used for removing the free fatty acids.

The Classical Alkaline Refining Method Usually Comprises the Following Steps

Step 1: Degumming with water to remove the easily hydratable phospholipids and metals.

Step 2: Addition of a small amount of phosphoric or citric acid to convert the remaining non-hydralable phospholipids (Ca, Mg salts) into hydratable phospholipids.

Step 3: Neutralising of the free fatty acids with a slight excess of sodium hydroxide solution, followed by the washing out of soaps and hydrated phospholipids.

Step 4: Bleaching with natural or acid-activated clay minerals to adsorb colouring components and to decompose hydroperoxides.

Step 5: Deodorising to remove volatile components, mainly aldehydes and ketones, with low threshold values for detection by taste or smell. Deodorisation is essentially a steam distillation process carried out at low pressures (2-6 mbar) and elevated temperatures (180-220°C).

For some oils, such as sunflower oil or rice bran oil, a clear table product is obtained by a dewaxing step or crystallization of the wax esters at low temperature, followed by filtration or centrifugation.

The alkaline neutralisation process has major drawbacks, the yield is relatively low and oil losses occur due to emulsification and saponification of neutral oil. Also, a considerable amount of liquid emuent is generated. The soaps are generally split with sulphuric acid to recover free fatty acids along with sodium sulphate and some fat-containing acid water steam.

In physical refining, the fatty acids are removed by a steam distillation (stripping) process similar to deodorisation. The low volatility of fatty acids (depending upon chain length) requires higher temperatures in physical refining than those required for only deodorisation. In practice, a maximum temperature of 240-250°C is sufficient to reduce the free fatty acid content to levels of about 0.05-0.1 percent. A prerequisite for physical refining is that phosphatides be removed to a level below 5 mg phosphorus/kg oil. In the classic refining process, this level is easily achieved during the neutralisation stage, but special degumming processes may be required for physical refining of high-phosphatide seed oils. These procedures rely on improved hydration of phospholipids by intimate contact of the oil with an aqueous solution of citric acid, phosphoric acid andlor sodium hydroxide, followed by bleaching.

It is unlikely that the mild reaction conditions during degumming and neutralisation will induce any significant undesirable changes in the oil composition. On the contrary, several impurities including oxidised components, trace metals and colouring materials are partially removed by entrainment with the phospholipids and soapstock. These impurities are further reduced during bleaching. Neutralisation also contributes significantly to the removal of contaminants such as aflatoxin and organophosphorous pesticides. Organochlorine pesticides and polycyclic aromatic hydrocarbons, if present, must be removed during the deodorisation/stripping stage and by active carbon treatment. Some loss of tocopherols and sterols during alkaline neutralisation usually occurs, however, under wellcontrolled conditions (minimising air contact) this need not exceed 5-10 percent.

Potential Side Reactions During High-temperature Processing

The possibility of negative effects of high temperatures during deodorisation and stripping has evoked concern. In several studies, extreme conditions of temperature and time (even with free access of air) were applied to produce significant quantitative results. However, the results of model studies should be related to practical processing conditions. As early as 1967-1979, the German Society for Fat Research (DGF) defined upper limits for deodorisation conditions [240°C for 2hrs, 270°C for 30 min.]

Good manufacturing practice also implies the following: the use of stainless steel equipment; the careful deaeration at < 100°C before heating to the final stripping temperature; the use of oxygen-free steam; and strict feedstock specifications (typically: Fe 0.1, Cu 0.01, P S, bleaching earth 5 mg/kg oil max.).

Investigations in which oil was maltreated under extreme conditions determined the effects of temperature (240-300°C) and time (30-180 min.) during "physical refining" of soybean oil (degummed with phosphoric acid and lightly bleached, but still containing 20 mg P, 0.35 mg Fe and 0.05 mg Cu per kg oil). A strong effect of temperature on the formation of trans fatty acids and polymeric compounds. Time also has a significant effect.

At 280-300°C, there was evidence for appreciable inter- or intra-esterification (increase in the content of saturated fatty acids at the 2-position of the triacylglycerols); substantial amounts of conjugated fatty acids were formed as well. The shaded areas indicate the usual range of processing conditions required for physical refining (270°C for 30 mini 250°C for 1 h; 240°C for 2 h; 220°C for 3 h). Under these conditions, all changes induced by the high temperature treatment appear to be relatively minor.

Cis-trans isomerisation. One of the most sensitive parametres used to detect chemical changes resulting from severe processing conditions is cis-trans isomerisation, especially of linolenic acid. The most complete study in this area was made by Eder (1982), who investigated the formation of geometric isomers in various oils on the laboratory, pilot plant and production scales. On a laboratory scale, with unbleached soyabean oil at 240°C, the formation of C_{18}:3 isomers (determined by GLC) was insignificant (less than 1 percent, even after 5 h, against 3 percent at 260°C).

The total trans contents (determined by infrared spectroscopy) for several types of oils deodorised in stainless steel equipment on the pilot plant scale. The data show a strong effect of temperature, especially between 240 and 270°C, and confirm the general trend. That is, even for highly unsaturated oils containing linolenic acid, the formation of trans isomers is slow under the conditions recommended for industrial deodorisation/physical refining (for example, 250°C max.). In conclusion, in the temperature range of 240-250°C, the amount of trans fatty acids formed from unsaturated oils is about 1 percent or less per hour. This is consistent with the data discussed earlier. Heating of oils in air, under simulated frying conditions, has led to similar types of isomerised fatty acids.

The formation of positional isomers (that is, double bonds shifted along the fatty acid chain) of linolenic and linoleic acid under deodorisation/physical refining conditions has not been reported.

Dimensahon and polymensahon. Quantitative data on the formation of polymeric compounds from bleached soybean oil are included. Up to about 260°C, their rate of formation appears to be low; above 260°C, the increase is more rapid. Similar trends were observed by Eder (1982). A rapid increase of the amount of polymeric triacylglycerols was only observed at 270°C. Deodorisation of soybean oil on a commercial scale (2 x 51 min at 240°C) resulted in an increase of 0.5 percent to 0.8 percent polymeric triacylglycerols.

When the temperature was increased to 270°C in one of the deodoriser trays, 1.5 percent polymers was found. This suggests that the content of di- and polymeric triacylglycerols do not normally exceed 1 percent by weight in properly refined oils and fats.

Strauss, Piater and Sterner (1982) carried out toxicological studies on mice fed concentrates (24 or 96 percent) of dimeric (incl. polymeric) triacylglycerols isolated from soybean oil that had been deodorised for 3.5 h at 220°C and 1 h at 270°C and contained 1.5 wt percent of dimers. The acute toxicity was found to be low with an LD 50 above 18 g/kg body weight. Long term administration of the dimer concentrate (12 months, 15 wt percent of the 24 percent concentrate in the diet) did not reveal any significant differences compared with the control group. The absorption of dimeric fatty acids was found to be poor. Therefore, the presence of small amounts of dimers and polymers in processed oils does not seem to present a physiological problem.

Physical Losses

During deodorisation or physical refining, volatile components are removed from the oil by the combination of high temperature, low pressure and stripping action of inert gas (steam). The degree of removal depends on the physical properties of the components (especially vapour pressure) and on the temperature and volume of steam passed through the oil. Some physical losses are highly desirable, for example, the removal of off- flavours, pesticides and polycyclic aromatic hydrocarbons, if present. Other losses of nutritionally valuable components, such as tocopherols and sterols, are potentially undesirable.

Quantitative data on the composition of deodoriser distillates and the removal of various types of contaminants from oils during deodorisation were reported. During high temperature deodorisation or physical refining, especially, the concentrations of organochlorine pesticides (lindane, DDT etc.), if present in the bleaching oil, are reduced to very low levels.

Polycyclic aromatic hydrocarbons (PAM) have been a reason for concern since they were detected in some types of crude oils and fats in the 1960s. An example of this is coconut oil from copra dried with non-purified smoke gases. Depending upon the number of aromatic rings, polycyclic aromatic hydrocarbons are classified as either light (3-4 rings) or heavy (5 or more rings). Some of these compounds have proven carcinogenic properties, for example, benz-a -pyrene. Light polycyclic aromatic hydrocarbons can be removed by deodorisation or physical refining, whereas heavy polycyclic aromatic hydrocarbons can be removed only by adsorption onto activated carbon. This treatment, which can be combined with the bleaching process, is effective in reducing the amount of polycyclic aromatic hydrocarbons to acceptable levels.

Some loss of tocopherols and sterols by evaporation during high temperature deodorisation and physical refining is inevitable. However, they have higher molecular masses and lower volatilities than free fatty acids and light polycyclic aromatic hydrocarbons. Consequently, the losses of tocopherols and sterols need not be severe if the processing conditions are well-chosen. Extreme conditions applied in some model studies do induce more drastic effects. Typical results obtained for soybean oil. After 2 hrs at 300°C (a drastic treatment), the tocopherols and sterols almost completely disappeared, whereas the actual reduction during physical refining at 240°C for 120 min is only about 15-20 percent. The total refining losses (including pretreatment) are about 25-35 percent. Many investigators report similar data for a variety of oils. The losses tend to be higher in physical refining than in alkaline refining because of higher stripping temperatures.

Under extreme conditions, some isomerisation of b -sitosterol can occur and individual tocopherols (a ,b , t , d) and sterols may behave differently when exposed to high temperatures. However, these phenomena are of minor significance under more realistic conditions. The percentage composition of the tocopherol and sterol fractions remains essentially unchanged during processing.

For refined oils such as sunflowerseed, cottonseed and rapeseed oil, an upper limit of 3035 percent loss of tocopherols throughout the entire processing would still satisfy the generally accepted criterion of oc-tocopherol equivalents/linoleic acid ³ 0.6 mg/g (eager, 1975). Soybean oil, which is high in g -tocopherol and consequently well-protected in vitro, is relatively low in a -tocopherol and cannot reach this ratio.

The, b -carotene in palm oil is another valuable component that needs consideration in the refining process. Special processes for retention are being designed. Olive oil and sesame oil are used in unrefined form since a specific taste is expected by consumers.

Fat Modification Processes

Hydrogenation. Hydrogenation of edible oils and fats has been applied on a large scale since the beginning of this century. The process is carried out in a three-phase system (hydrogen gas, liquid oil and solid catalyst) at temperatures ranging from about 120°C to about 220°C max. in the final stages of the reaction. The catalyst consists of small nickel crystallites supported by an inorganic oxide, usually silica or alumina. After the reaction, the catalyst is filtered off and any traces of residual nickel are removed in post-refining to a level of about 0.1 mg/kg or below.

Hydrogenation is a series of consecutive reactions following pseudo first order reaction kinetics:

K_3	K_2	K_1	
18:3——>	18:2———>	18:1———>	18:0

in which K_3, K_2 and K_1 are the reaction rate constants for linolenic, linoleic and oleic acid, respectively. The K3 is much larger than K2 and K,. In nearly all hydrogenations, linolenic acid is transformed into compounds which are less saturated. Depending upon the reaction conditions, the socalled Selectivity Ratio (K_2/K_1) can change significantly; for example, for nickel catalysts it varies from approximately 10 at a low temperature to 50 or 100 at high temperature. A high Selectivity Ratio implies that relatively little saturated acid is formed and that monounsaturated fatty acids are the dominant reaction product.

Apart from reduction of unsaturation, isomerisation of double bonds also takes place during hydrogenation: geometric isomerisation (cis-trans) and positional isomerisation. The mechanisms of hydrogenation and isomerisation are strongly interrelated. Initially, a half-hydrogenated intermediate is formed in which the molecule adsorbed to the catalyst surface by a single bond can freely rotate. Addition of a second hydrogen atom results in saturation, whereas abstraction of a hydrogen atom from the half-hydrogenated intermediate to the nickel surface yields either the original molecule or a positional or geometric isomer. Hydrogenation of polyenoic fatty acids occurs, at least partially, through conjugated isomers (for example, c9, t11 or t10, c12) which are highly reactive and, therefore, rapidly converted into cis or trans monoenoic acids without accumulating.

Of the total amount of trans fatty acids present in hydrogenated oils and fats, the greater part by far are trans monoenes. Due to the important role hydrogenation plays in the production of plastic fats, trans fatty acids may be found in considerable amounts in many products. The amount of cis,trans and trans,cis dienes is much smaller, whilst the level of trans,trans dienes seldom exceeds 1 percent.

Apart from hydrogenation, there are two other important fat modification technologies. The first is interesterification, the random rearrangement of the fatty acids in the triglyceride molecule under the influence of a moderately alkaline catalyst. This modifies the melting behaviour of the fat without changing the nature of the fatty acids. The second is fractionation, the controlled separation of oil/fat fractions by low temperature (dry fractionation), or solvents (solvent fractionation). In this process, there are no changes in the chemical nature of the fatty acids. Palm oil is fractionated into palmolein and palmstearin.

Interrelationships of fat modification technologies. Industry uses various oils and fats interchangeably while retaining constant quality. Generally, the least expensive combination of raw materials compatible with the required quality is chosen. Hydrogenation greatly extends the number of fats available with specified melting behaviour and this increases interchangeability and lowers costs.

In non-hydrogenation situations, a combination of interesterification, fractionation and selection of the starting oil may be acceptable solutions for limiting the formation of isomers in terms of product quality, however, the costs are higher. While the actual specific modification, for example, hydrogenation or interesterification, is relatively inexpensive, the costs of losing flexibility can be significantly higher. The processes for oil modification may change as new compositions become available through plant biotechnology.

Other Considerations

Storage, transport and packaging of oils. Oils and fats must be protected against oxidative deterioration, contamination with water, dirt, or other fats, absorption of foreign odours and tastes, thermal deterioration, and entry of foreign substances from packaging and lining materials. Temperature, oxygen pressure, oxidation products, trace metals, oxidative and lipolytic enzymes, reduction in natural antioxidants and visible and ultraviolet light are all factors in such deterioration. The use of low storage temperatures, nitrogen or vacuum packaging; the avoidance of copper, copper alloys and iron as construction material of storage vessels; and the use of synthetic or natural antioxidants and metal sequestrants as additives, work to prevent deterioration of oil during storage.

Selection of oil processing technology. The use of oil processing technology and its application in products is influenced by several factors. For example, the demand for triacylglycerols with specific fatty acids in the 1, 2 and 3 positions of the molecule can be met through enzymatic interesterification which uses lipases as catalysts in the interesterification process. The more readily absorbed fatty acid in the 2-position can provide for specific triacylglycerols with medical benefits. For example, essential fatty acids can be supplied to patients with various types of fat malabsorption or energy can be given to infants with palmitic acid in the 2-position. Another is the increased use of physical refining because of consumer pressure for less "chemical" processing.

Hydrogenation

Hydrogenation – to treat with hydrogen – is a chemical reaction between molecular hydrogen (H_2) and another compound or element, usually in the presence of a catalyst such as nickel, palladium or platinum. The process is commonly employed to reduce or saturate organic compounds. Hydrogenation typically constitutes the addition of pairs of hydrogen atoms to a molecule, generally an alkene. Catalysts are required for the reaction to be usable; non-catalytic hydrogenation takes place only at very high temperatures. Hydrogenation reduces double and triple bonds in hydrocarbons.

Because of the importance of hydrogen, many related reactions have been developed for its use. Most hydrogenations use gaseous hydrogen (H_2), but some involve the alternative sources of hydrogen, not H_2: these processes are called transfer hydrogenations. The reverse reaction, removal of hydrogen from a molecule, is called dehydrogenation. A reaction where bonds are broken while hydrogen is added is called hydrogenolysis, a reaction that may occur to carbon-carbon and carbon-heteroatom (oxygen, nitrogen or halogen) bonds. Hydrogenation differs from protonation or hydride addition: in hydrogenation, the products have the same charge as the reactants.

Hydrogenation of unsaturated fats produces saturated fats. In the case of partial hydrogenation, trans fats may be generated as well.

Process

Hydrogenation has three components, the unsaturated substrate, the hydrogen (or hydrogen source) and, invariably, a catalyst. The reduction reaction is carried out at different temperatures and pressures depending upon the substrate and the activity of the catalyst.

Substrate

The addition of H_2 to an alk*e*ne affords an alk*a*ne in the prototypical reaction:

$$RCH{=}CH_2 + H_2 \rightarrow RCH_2CH_3 \text{ (R = alkyl, aryl)}$$

Hydrogenation is sensitive to steric hindrance explaining the selectivity for reaction with the exocyclic double bond but not the internal double bond.

An illustrative example of a hydrogenation reaction is the addition of hydrogen to maleic acid to form succinic acid. Numerous important applications of this petrochemical are found in pharmaceutical and food industries.

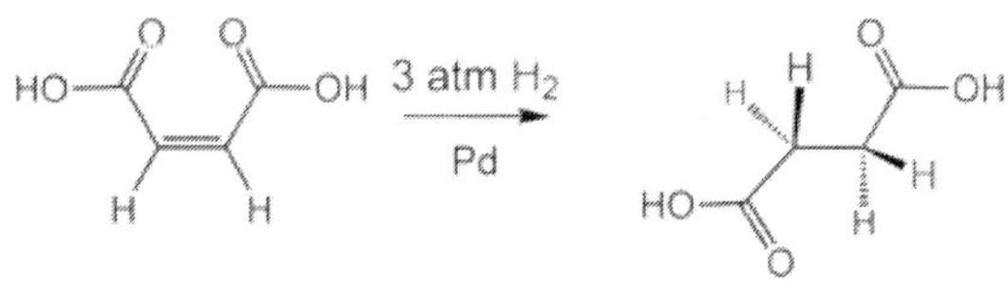

An important characteristic of alkene and alkyne hydrogenations, both the homogeneously and heterogeneously catalyzed versions, is that hydrogen addition occurs with "syn addition", with hydrogen entering from the least hindered side. Typical substrates are listed in the table

Substrates for and products of hydrogenation		
substrate	***product***	***comments***
alkene, $R_2C{=}CR'_2$	alkane, $R_2CHCHR'_2$	many catalysts, one application is margarine
alkyne, RCCR	alkene, cis-$RHC{=}CHR'$	over-hydrogenation to alkane can be problematic
aldehyde, RCHO	primary alcohol, RCH_2OH	easy substrate
ketone, R_2CO	secondary alcohol, R_2CHOH	more challenging than RCHO, prochiral for unsymmetrical ketones
ester, RCO_2R'	two alcohols, RCH_2OH + R'OH	challenging substrate
imine, RR'CNR"	amine, RR'CHNHR"	easy substrate, often use transfer hydrogenation, actual precursor is N-protonated
amide, $RC(O)NR'_2$	amine, $RCH_2NR'_2$	challenging substrate
nitrile, RCN	primary amine, RCH_2NH_2	product amine reactive toward precursor nitrile in some cases
nitro, RNO_2	amine, RNH_2	commercial applications use heterogeneous Ni and Ru catalysts; major application is aniline

Catalysts

With rare exceptions, no reaction below 480 °C (750 K or 900 °F) occurs between H_2 and organic compounds in the absence of metal catalysts. The unsaturated substrate is chemisorbed onto the catalyst, with most sites covered by the substrate. In heterogeneous catalysts, hydrogen forms surface hydrides (M-H) from which hydrogens can be transferred to the chemisorbed substrate. Platinum, palladium, rhodium, and ruthenium form highly active catalysts, which operate at lower temperatures and lower pressures of H_2. Non-precious metal catalysts, especially those based on nickel (such as Raney nickel and Urushibara nickel) have also been developed as economical alternatives, but they are often slower or require higher temperatures. The trade-off is activity (speed of reaction) vs. cost of the catalyst and cost of the apparatus required for use of high pressures. Notice that the Raney-nickel catalysed hydrogenations require high pressures:

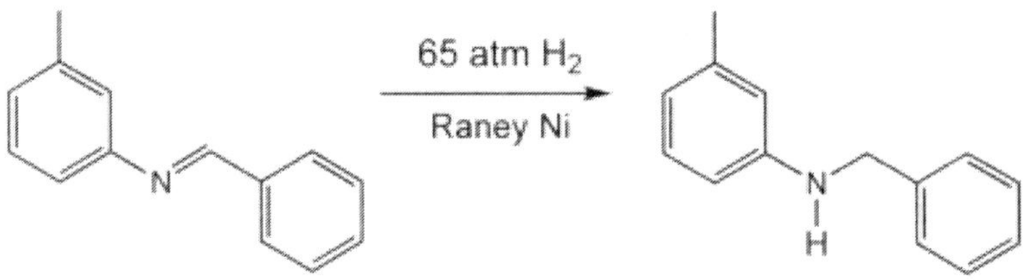

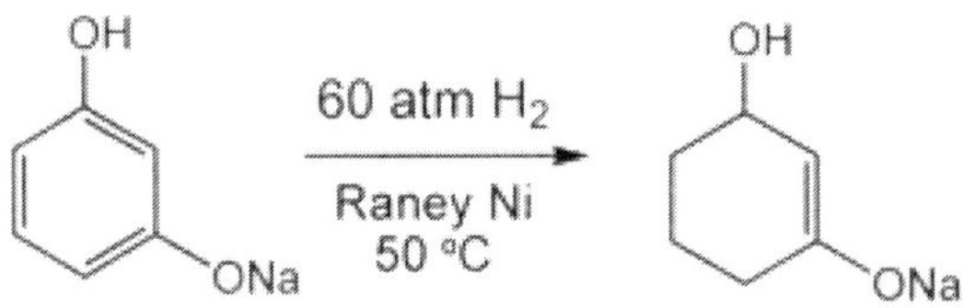

Two broad families of catalysts are known - homogeneous catalysts and heterogeneous catalysts. Homogeneous catalysts dissolve in the solvent that contains the unsaturated substrate. Heterogeneous catalysts are solids that are suspended in the same solvent with the substrate or are treated with gaseous substrate.

Homogeneous Catalysts

Illustrative homogeneous catalysts include the rhodium-based compound known as Wilkinson's catalyst and the iridium-based Crabtree's catalyst. An example is the hydrogenation of carvone:

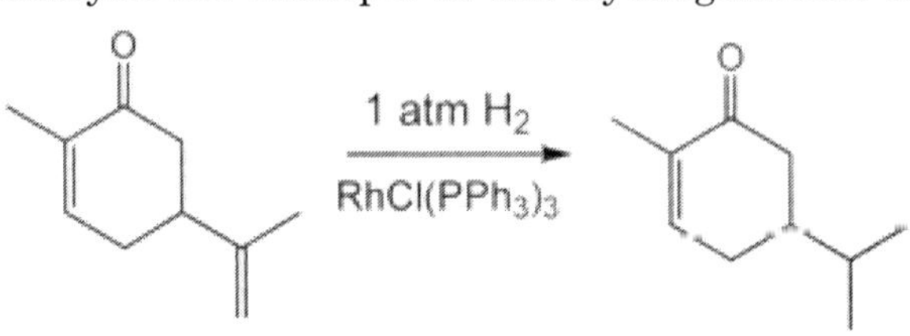

Hydrogenation is sensitive to steric hindrance explaining the selectivity for reaction with the exocyclic double bond but not the internal double bond.

The activity and selectivity of homogeneous catalysts is adjusted by changing the ligands. For prochiral substrates, the selectivity of the catalyst can be adjusted such that one enantiomeric product is favoured. Asymmetric hydrogenation is also possible via heterogeneous catalysis on a metal that is modified by a chiral ligand.

Heterogeneous Catalysts

Heterogeneous catalysts for hydrogenation are more common industrially. As in homogeneous catalysts, the activity is adjusted through changes in the environment around the metal, i.e. the coordination sphere. Different faces of a crystalline heterogeneous catalyst display distinct activities, for example. Similarly, heterogeneous catalysts are affected by their supports, i.e. the material upon with the heterogeneous catalyst is bound.

In many cases, highly empirical modifications involve selective "poisons". Thus, a carefully chosen catalyst can be used to hydrogenate some functional groups without affecting others, such as the hydrogenation of alkenes without touching aromatic rings, or the selective hydrogenation of alkynes to alkenes using Lindlar's catalyst. For example, when the catalyst palladium is placed on barium sulfate and then treated with quinoline, the resulting catalyst reduces alkynes only as far as alkenes. The Lindlar catalyst has been applied to the conversion of phenylacetylene to styrene.

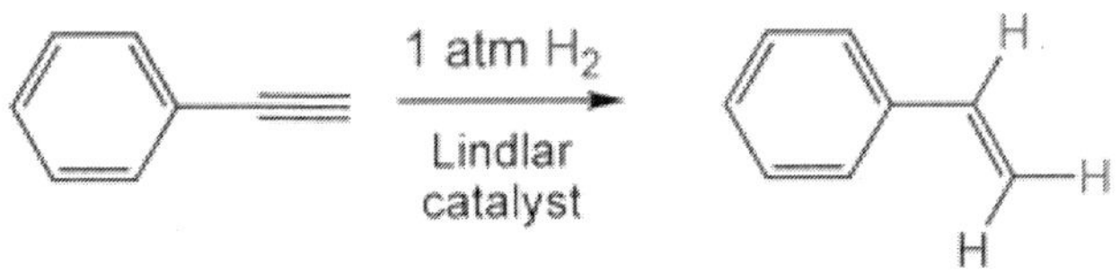

Asymmetric hydrogenation is also possible via heterogeneous catalysis on a metal that is modified by a chiral ligand.

Hydrogen Sources

For hydrogenation, the obvious source of hydrogen is H_2 gas itself, which is typically available commercially within the storage medium of a pressurized cylinder. The hydrogenation process often uses greater than 1 atmosphere of H_2, usually conveyed from the cylinders and sometimes augmented by "booster pumps". Gaseous hydrogen is produced industrially from hydrocarbons by the process known as steam reforming.

Transfer Hydrogenation

Hydrogen also can be extracted ("transferred") from "hydrogen-donors" in place of H_2 gas. Hydrogen donors, which often serve as solvents include hydrazine, dihydronaphthalene, dihydroanthracene, isopropanol, and formic acid.

In organic synthesis, transfer hydrogenation is useful for the asymmetric reduction of polar unsaturated substrates, such as ketones, aldehydes, and imines.

Electrolytic Hydrogenation

Polar substrates such as nitriles can be hydrogenated electrochemically, using protic solvents and reducing equivalents as the source of hydrogen.

Thermodynamics and Mechanism

Hydrogenation is a strongly exothermic reaction. In the hydrogenation of vegetable oils and fatty acids, for example, the heat released is about 25 kcal per mole (105 kJ/mol), sufficient to raise the temperature of the oil by 1.6–1.7 °C per iodine number drop. The mechanism of metal-catalyzed hydrogenation of alkenes and alkynes has been extensively studied. First of all isotope labelling using deuterium confirms the regiochemistry of the addition:

$$RCH{=}CH_2 + D_2 \rightarrow RCHDCH_2D$$

Heterogeneous Catalysis

Steps in the hydrogenation of a C=C double bond at a catalyst surface, for example Ni or Pt :

(1) The reactants are adsorbed on the catalyst surface and H_2 dissociates.

(2) An H atom bonds to one C atom. The other C atom is still attached to the surface.

(3) A second C atom bonds to an H atom. The molecule leaves the surface.

On solids, the accepted mechanism is the Horiuti-Polanyi mechanism:

1. Binding of the unsaturated bond, and hydrogen dissociation into atomic hydrogen onto the catalyst
2. Addition of one atom of hydrogen; this step is reversible
3. Addition of the second atom; effectively irreversible under hydrogenating conditions.

In the second step, the metallointermediate formed is a saturated compound that can rotate and then break down, again detaching the alkene from the catalyst.

Consequently, contact with a hydrogenation catalyst necessarily causes *cis-trans*-isomerization, because the isomerization is thermodynamically favourable. This is a problem in partial hydrogenation, while in complete hydrogenation the produced *trans*-alkene is eventually hydrogenated.

For aromatic substrates, the first bond is hardest to hydrogenate because of the free energy penalty for breaking the aromatic system. The product of this is a cyclohexadiene, which is extremely active and cannot be isolated; in conditions reducing enough to break the aromatization, it is immediately reduced to a cyclohexene.

The cyclohexene is ordinarily reduced immediately to a fully saturated cyclohexane, but special modifications to the catalysts (such as the use of the anti-solvent water on ruthenium) can preserve some of the cyclohexene, if that is a desired product.

Homogeneous Catalysis

In many homogeneous hydrogenation processes, the metal binds to both components to give an intermediate alkene-metal$(H)_2$ complex. The general sequence of reactions is assumed to be as follows or a related sequence of steps:

- binding of the hydrogen to give a dihydride complex ("oxidative addition"):

$$L_nM + H_2 \rightarrow L_nMH_2$$

- binding of alkene:

$$L_nM(\eta^2H_2) + CH_2{=}CHR \rightarrow L_{n-1}MH_2(CH_2{=}CHR) + L$$

- transfer of one hydrogen atom from the metal to carbon (migratory insertion)

$$L_{n-1}MH_2(CH_2{=}CHR) \rightarrow L_{n-1}M(H)(CH_2\text{-}CH_2R)$$

- transfer of the second hydrogen atom from the metal to the alkyl group with simultaneous dissociation of the alkane ("reductive elimination")

$$L_{n-1}M(H)(CH_2\text{-}CH_2R) \rightarrow L_{n-1}M + CH_3\text{-}CH_2R$$

Preceding the oxidative addition of H_2 is the formation of a dihydrogen complex.

Inorganic Substrates

The hydrogenation of nitrogen to give ammonia is conducted on a vast scale by the Haber-Bosch process, consuming an estimated 1% of the world's energy supply.

$$N{\equiv}N \xrightarrow[\text{Fe catalyst, } 350\text{-}550\ ^\circ C]{200 \text{ atm } H_2} 2\,NH_3$$

Oxygen can be partially hydrogenated to give hydrogen peroxide, although this process has not been commercialized.

Industrial Applications

Catalytic hydrogenation has diverse industrial uses. Most frequently, industrial hydrogenation relies on heterogeneous catalysts. In petrochemical processes, hydrogenation is used to convert alkenes and aromatics into saturated alkanes (paraffins) and cycloalkanes (naphthenes), which are less toxic and less reactive.

For example, mineral turpentine is usually hydrogenated. Hydrocracking of heavy residues into diesel is another application. In isomerization and catalytic reforming processes, some hydrogen pressure is maintained to hydrogenolyze coke formed on the catalyst and prevent its accumulation.

Hydrogenation is a useful reaction for converting more oxidized oxygen and nitrogen compounds such as aldehydes, imines and nitriles into the corresponding saturated compounds, i.e. alcohols and amines. Primary alcohols can be synthesized from aldehydes by hydrogenation. Thus, alkyl aldehydes, which can be synthesized with the oxo process from carbon monoxide and an alkene, can be converted to alcohols. E.g. 1-propanol is produced from propionaldehyde, produced from ethene and carbon monoxide.

Xylitol, a polyol, is produced by hydrogenation of the sugar xylose, an aldehyde. Primary amines can be synthesized by hydrogenation of nitriles, while nitriles are readily synthesized from cyanide and a suitable electrophile.

For example, isophorone diamine, a precursor to the polyurethane monomer isophorone diisocyanate, is produced from isophorone nitrile by a tandem nitrile hydrogenation/reductive amination by ammonia, wherein hydrogenation converts both the nitrile into an amine and the imine formed from the aldehyde and ammonia into another amine.

In the Food Industry

The largest scale application of hydrogenation is for the processing of vegetable oils.

Typical vegetable oils are derived from polyunsaturated fatty acids (containing more than one carbon-carbon double bonds). Their partial hydrogenation reduces most but not all, of these carbon-carbon double bonds.

The degree of hydrogenation is controlled by restricting the amount of hydrogen, reaction temperature and time, and the catalyst.

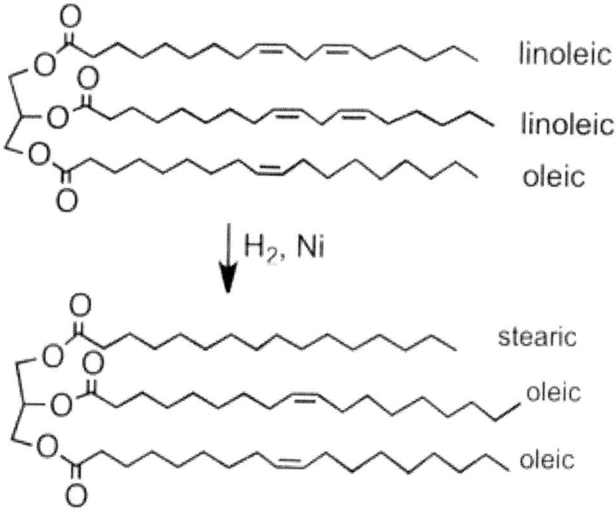

Partial hydrogenation of a typical plant oil to a typical component of margarine. Most of the C=C double bonds are removed in this process, which elevates the melting point of the product.

Hydrogenation converts liquid vegetable oils into solid or semi-solid fats, such as those present in margarine. Changing the degree of saturation of the fat changes some important physical properties such as the melting range, which is why liquid oils become semi-solid. Solid or semi-solid fats are preferred for baking because the way the fat mixes with flour produces a more desirable texture in the baked product.

Because partially hydrogenated vegetable oils are cheaper than animal source fats, are available in a wide range of consistencies, and have other desirable characteristics (e.g., increased oxidative stability/longer shelf life), they are the predominant fats used as shortening in most commercial baked goods.

Health Implications

A side effect of incomplete hydrogenation having implications for human health is the isomerization of some of the remaining unsaturated carbon bonds. The cis configuration of these double bonds predominates in the unprocessed fats in most edible fat sources, but incomplete hydrogenation partially converts these molecules to trans isomers, which have been implicated in circulatory diseases including heart disease.

The conversion from cis to trans bonds is favoured because the trans configuration has lower energy than the natural cis one. At equilibrium, the trans/cis isomer ratio is about 2:1. Food legislation in the US and codes of practice in EU have long required labels declaring the fat content of foods in retail trade and, more recently, have also required declaration of the trans fat content. The use of trans fats in human food products has been effectively banned in Denmark (since 2003) and Switzerland (2008).

In the US, local legislation banned trans fats from restaurants and public kitchens in New York City (since 2005) and California. Other countries and regions have introduced mandatory labelling of trans fats on food products and appealed to the industry for voluntary reductions.

Partially hydrogenated fats can be replaced, for instance by interesterified fats or by formulation such as adding palm oil to unsaturate-rich oils.

History

Heterogeneous Catalytic Hydrogenation

The earliest hydrogenation is that of platinum catalyzed addition of hydrogen to oxygen in the Döbereiner's lamp, a device commercialized as early as 1823. The French chemist Paul Sabatier is considered the father of the hydrogenation process.

In 1897, building on the earlier work of James Boyce, an American chemist working in the manufacture of soap products, he discovered that the introduction of a trace of nickel as a catalyst facilitated the addition of hydrogen to molecules of gaseous hydrocarbons in what is now known as the Sabatier process.

For this work Sabatier shared the 1912 Nobel Prize in Chemistry. Wilhelm Normann was awarded a patent in Germany in 1902 and in Britain in 1903 for the hydrogenation of liquid oils, which was the beginning of what is now a world wide industry. The commercially important Haber–Bosch process, first described in 1905, involves hydrogenation of nitrogen.

In the Fischer–Tropsch process, reported in 1922 carbon monoxide, which is easily derived from coal, is hydrogenated to liquid fuels.

Also in 1922, Voorhees and Adams described an apparatus for performing hydrogenation under pressures above one atmosphere. The Parr shaker, the first product to allow hydrogenation using elevated pressures and temperatures, was commercialized in 1926 based on Voorhees and Adams' research and remains in widespread use.

In 1924 Murray Raney developed a nickel fine powder catalyst named after him which is still widely used in hydrogenation reactions such as conversion of nitriles to amines or the production of margarine.

Homogeneous Catalytic Hydrogenation

The history of homogeneous hydrogenation has been assigned to the development the Meerwein–Ponndorf–Verley reduction of ketones using aluminium alkoxides. In the 1930s, Calvin discovered that copper(II) complexes oxidized H_2.

The 1960s witnessed the development of well defined homogeneous catalysts using transition metal complexes, e.g., Wilkinson's catalyst (RhCl(PPh3)3). Soon thereafter Schrock and Obsorne's discovered that cationic Rh and Ir catalyze the hydrogenation of alkenes and carbonyls. In the 1970s, asymmetric hydrogenation was demonstrated in the synthesis of L-DOPA and the 1990s saw the invention of Noyori asymmetric hydrogenation.

The development of homogeneous hydrogenation was influenced by work started in the 1930s and 1940s on the oxo process and Ziegler–Natta polymerization. More recently, homogeneous hydrogenation has been used to promote reductive C-C bond formation.

Metal-free Hydrogenation

For most practical purposes, hydrogenation requires a metal catalyst. Hydrogenation can, however, proceed from some hydrogen donors without catalysts, illustrative hydrogen donors being diimide and aluminium isopropoxide.

Some metal-free catalytic systems have been investigated in academic research. One such system for reduction of ketones consists of tert-butanol and potassium tert-butoxide and very high temperatures. The reaction depicted below describes the hydrogenation of benzophenone:

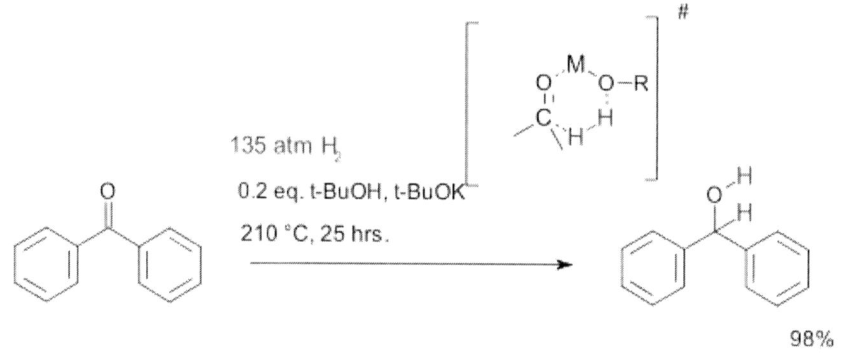

A chemical kinetics study found this reaction is first-order in all three reactants suggesting a cyclic 6-membered transition state. Another system for metal-free hydrogenation is based on the phosphine-borane, compound 1, which has been called a *frustrated Lewis pair.*

It reversibly accepts dihydrogen at relatively low temperatures to form the phosphonium borate 2 which can reduce simple hindered imines.

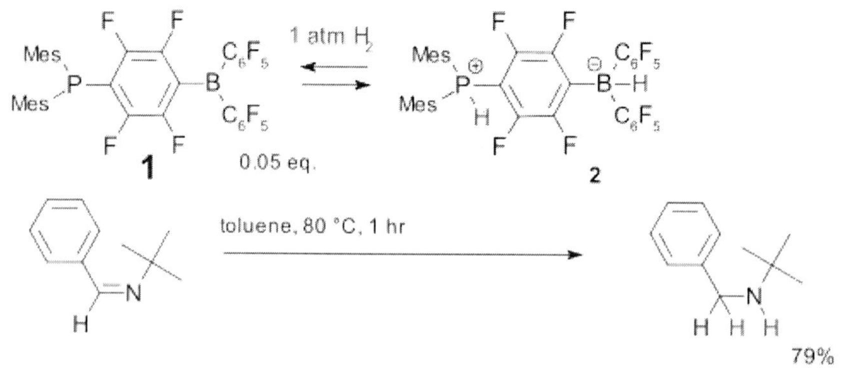

The reduction of nitrobenzene to aniline has been reported to be catalysed by fullerene, its mono-anion, atmospheric hydrogen and UV light.

Equipment Used for Hydrogenation

Today's bench chemist has three main choices of hydrogenation equipment:

- Batch hydrogenation under atmospheric conditions
- Batch hydrogenation at elevated temperature and/or pressure
- Flow hydrogenation

Batch Hydrogenation Under Atmospheric Conditions

The original and still a commonly practised form of hydrogenation in teaching laboratories, this process is usually effected by adding solid catalyst to a round bottom flask of dissolved reactant which has been evacuated using nitrogen or argon gas and sealing the mixture with a penetrable rubber seal. Hydrogen gas is then supplied from a H_2-filled balloon.

The resulting three phase mixture is agitated to promote mixing. Hydrogen uptake can be monitored, which can be useful for monitoring progress of a hydrogenation. This is achieved by either using a graduated tube containing a coloured liquid, usually aqueous copper sulfate or with gauges for each reaction vessel.

Batch Hydrogenation at Elevated Temperature And/or Pressure

Since many hydrogenation reactions such as hydrogenolysis of protecting groups and the reduction of aromatic systems proceed extremely sluggishly at atmospheric temperature and pressure, pressurised systems are popular.

In these cases, catalyst is added to a solution of reactant under an inert atmosphere in a pressure vessel. Hydrogen is added directly from a cylinder or built in laboratory hydrogen source, and the pressurized slurry is mechanically rocked to provide agitation, or a spinning basket is used. Heat may also be used, as the pressure compensates for the associated reduction in gas solubility.

Flow Hydrogenation

Flow hydrogenation has become a popular technique at the bench and increasingly the process scale. This technique involves continuously flowing a dilute stream of dissolved reactant over a fixed bed catalyst in the presence of hydrogen. Using established HPLC technology, this technique allows the application of pressures from atmospheric to 1,450 psi (100 bar). Elevated temperatures may also be used. At the bench scale, systems use a range of pre-packed catalysts which eliminates the need for weighing and filtering pyrophoric catalysts.

Industrial Reactors

Catalytic hydrogenation is done in a tubular plug-flow reactor (PFR) packed with a supported catalyst. The pressures and temperatures are typically high, although this depends on the catalyst. Catalyst loading is typically much lower than in laboratory batch hydrogenation, and various promoters are added to the metal, or mixed metals are used, to improve activity, selectivity and catalyst stability. The use of nickel is common despite its low activity, due to its low cost compared to precious metals. Gas Liquid Induction Reactors (Hydrogenator) are also used for carrying out catalytic hydrogenation.

Interesterification

Interesterification can be defined as a redistribution of the fatty acid moieties present in a triglyceride oil over its glycerol moieties. This can be random when the reaction is carried out in a single, liquid phase. When during the reaction higher melting triglycerides are allowed to crystallise, their mainly saturated fatty acids are withdrawn from the liquid phase so that the distribution of the fatty acids still present in the liquid phase will no longer be random.

It will continue to redistribute these residual fatty acids and thereby form further higher melting triglycerides that then also crystallize. This is called 'directed interesterification'. Directed interesterification processes are no longer practised industrially.

Early processes that arrived at a randomised fatty acid distribution did so by a two-step process. Grün allowed triglycerides to react with glycerol to generate partial glycerides that in a second step, were esterified with butyric acid and capronic acid to arrive at a synthetic butter fat. Normann went the opposite way and acidolysed the fat with butyric acid and then reacted the free fatty acids present with glycerol. Both these processes required a catalyst and were operated at elevated temperatures (140 to 250°C). A high temperature process using water to partially hydrolyse the triglycerides and recombine them when the water was flashed off has been described by Eckey.

He also introduced the low temperature (< 100°C) interesterification process using sodium methanolate (or sodium methoxide, $NaOCH_3$). Alkali metals such as sodium and potassium and their liquid alloy (NaK) were used in the US for the directed interesterification of lard yielding a shortening product with extended plastic range. In Europe, various hardstocks were produced by interesterification for margarine with a high linoleic acid content. One such margarine [5] consists of 15-30% of a fraction of a fully hydrogenated interesterification mixture of palm kernel oil and palm oil and 70-85% liquid oil.

Interesterified fat is a type of oil where the fatty acids have been moved from one triglyceride molecule to another. This is generally done to modify the melting point, slow rancidification and create an oil more suitable for deep frying or making margarine with good taste and low saturated fat content. It is not the same as partial hydrogenation which produces trans fatty acids, but interesterified fats used in the food industry can come from hydrogenated fat, for simplicity and economic reasons.

Chemistry

Fats such as soybean oil consist mainly of various triglycerides which are made up of a glycerol backbone esterified to three fatty acid molecules. The triglycerides contain a mixture of saturated, monounsaturated and polyunsaturated fatty acids. Interesterification is carried out by blending the desired oils and then rearranging the fatty acids over the glycerol backbone with, for instance, the help of catalysts or lipase enzymes. Polyunsaturated fatty acids (PUFAs) decrease the melting point of fats significantly.

A triglyceride containing three saturated fatty acids is generally solid at room temperature and not very desirable for many applications. Rearranging these triglycerides with oils containing unsaturated fatty acids lowers the melting point and creates fats with properties better suited for target food products.

In addition, blending interesterified oils with liquid oils allows the reduction in saturated fatty acids in many trans fatty acid free food products. The interesterified fats can be separated through controlled crystallization, also called fractionation.

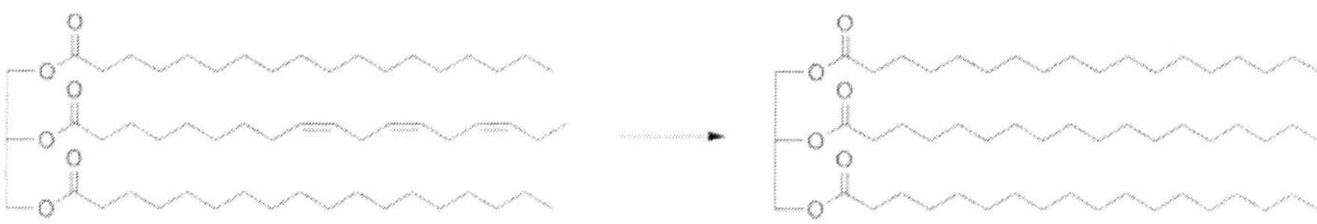

Figure: *A triglyceride with a PUFA (linolenic acid) at the sn-2 position which is replaced by a saturated fatty acid (stearic acid) via interesterification.*

In vegetable polyunsaturated oils, the PUFA is commonly found at the middle position (sn2) on the glycerol. Stearic acid is not usually found at sn2 in vegetable oils used in the human diet.

Hypotheses

In most vegetable dietary fats, palmitic (C16:0) and stearic acids (C18:0) mainly occupy the 1- and 3-positions of the triacylglycerol molecule, whereas an unsaturated fatty acid such as oleic acid or linoleic acid (18:2 cis,cis-9,12) usually occupies the 2-position. In animal fats, this is not the case. Interesterification of vegetable oils will enhance the amount of saturated fatty acids at the 2-position. Fatty acids at the 2-position are biologically different from fatty acids at the 1 and 3 position because they are handled differently during digestion and metabolism, and a relevant scientific question is whether there are health effects following from this. Although this question has received relatively little attention in dietary fats and health research, there are a number of good controlled human intervention studies that have addressed it.

In studies addressing the health effects of interesterification as such, a diet high in interesterified fat should be compared with a diet high in a noninteresterified fat with the same fatty acid composition. If the two diets show similar changes in the resulting blood lipid profiles (i.e. not different from each other), this indicates interesterification has no effect on metabolism or biological effects. Conversely, effects of interesterification cannot be properly addressed if the interesterified fat and the noninteresterified fat being compared have different fatty acid compositions.

Individual Studies

Zock et al. compared the effects of an IE test fat with 40% C16:0 on the 2-position with a noninteresterified test fat with only 6.5% C16:0 on the 2-position in a 3-week diet study. Despite the very high intakes and the marked difference in positional distribution, no statistically significant effects on fasting blood lipids were observed in the group as a whole. Nestel et al. examined the effects of an IE fat blend with 25% C16:0 on the 2-position with a native fat blend with only 9% C16:0 on the 2-position.

Again, despite a high intake level and the clear difference in positional distribution of the fats fed, no effects were observed on fasting blood lipids. Meijer and Weststrate examined the effects of interesterification, using a 'real' hardstock as applied in foods. The control was the same fat blend with a similar fatty acid composition, but not interesterified. The IE fat blend contained more C16:0 on the 2-position (18%) than the control blend (7%). None of the fasting levels of blood lipids measured after 3 weeks showed any change related to treatment of the fat blend. Fasting glucose level was also not affected.

In 1970, Grande et al. used interesterification to prepare a blend of fats and oils mimicking the fatty acid composition of cocoa butter. No difference between the interesterified fat blend and cocoa butter was observed in levels of total cholesterol in fasting blood.

Recently, in a study funded by the Malaysian Palm Oil Board, Sundram et al. compared the effects of three types of fat: native palm olein, a blend with partially hydrogenated soybean oil and an interesterified mixture of oils. They concluded both the IE blend and the partially hydrogenated fat blend increased the fasting LDL/HDL-cholesterol ratio, indicating an adverse effect on CVD risk. Sundram et al. also found that fasting plasma glucose levels were higher after 4 weeks on the interesterified fat than after the other diets. For the postprandial study the glucose incremental area under the curve (IAUC) following the IE meal was 40% greater than after either other meal ($p<0.001$), and was linked to relatively depressed insulin and C-peptide ($p<0.05$).

As was pointed out in a letter to the Editor by Destaillats et al., a major limitation of the Sundram study is that the diets differed in overall fatty acid composition. The interesterified fat had 30% more saturated and 57% less monounsaturated fatty acids than the untreated palm olein. The direction of the effects on blood lipids are in line with what can be predicted based on these differences in fatty acid content between the study diets (Mensink 2003).

Another recent study by Berry et al. compared shea butter (3% C18:0 on the 2-position) and interesterified shea butter (23% C18:0 on the 2-position), while keeping overall fatty acid composition of the diets constant. This study found no effects of interesterification on fasting levels of blood lipids, glucose and insulin. This is line with a number of other human intervention studies.

Christophe et al. have studied the effect of interesterification of butter oil. In a small pilot study, they observed an 11% lower blood total cholesterol level after interesterification. However, in a larger, better designed study the same authors could not reproduce the cholesterol-lowering effects.

Interesterification Mechanisms

The methanolate being an alcoholate, it did not require much imagination to suggest that the catalytically active intermediate in the sodium methanolate catalysed interesterification was indeed a glycerolate. Around the same time, another mechanism was proposed involving an enolate resulting from the α-hydrogen abstraction by the methanolate anion from a fatty acid moiety.

This mechanism assumed a β-keto ester to be the catalytically active intermediate but the observation that the concentration of this ester increased gradually during an interesterification reaction made this mechanism much less likely. Much later Liu demonstrated that the presence of this α hydrogen was essential for the interesterification to proceed. Dijkstra proposed a mechanism that takes all the observations known to him at that time into account and provides a tentative explanation for observations that had not been adequately explained by earlier mechanism proposals.

According to this mechanism, the sodium methanolate that is added to the oil to be interesterified can react in various ways giving rise to a number of different anions, one of which is a glycerolate anion. However, if the concentration of free hydroxyl groups in the oil is low, enol formation according to:

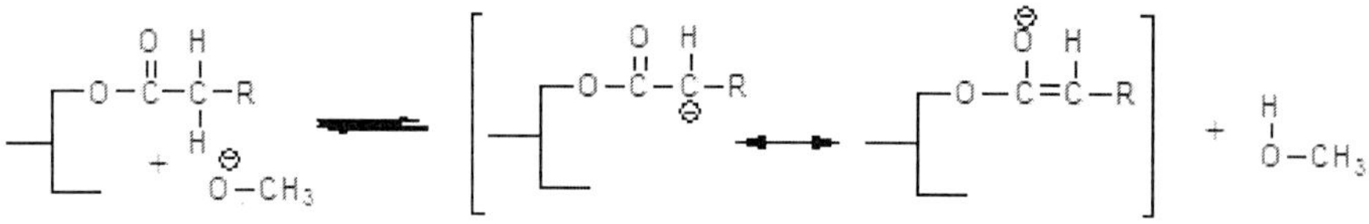

is probably the most likely reaction. This reaction liberates methanol and just like methanol that has been added to the reaction mixture this methanol can react to form a fatty acid methyl ester and a glycerolate anion according to:

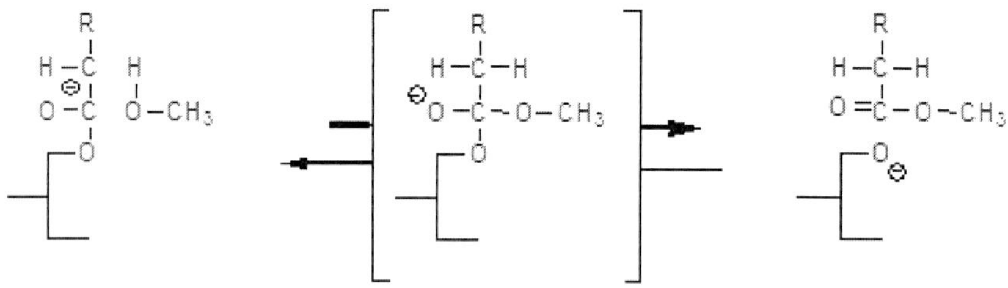

This reaction explains that interesterification reaction mixtures contain an amount of fatty acid methyl esters (FAME) that is equivalent to the amount of sodium methoxide added. It also explains that further methanol added to the reaction mixture eventually ends up as FAME. Perhaps some FAME are also formed on addition of the sodium methoxide according to:

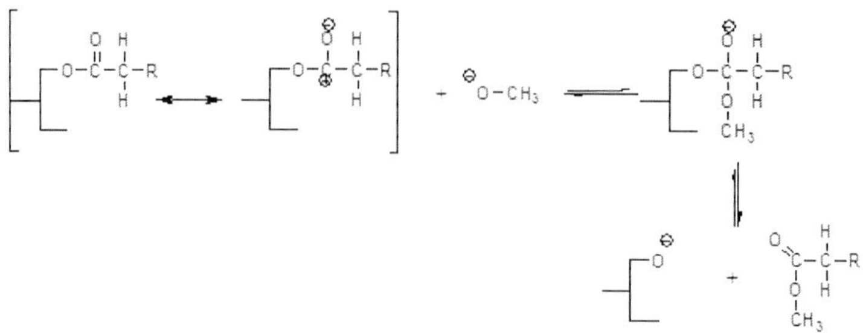

The above reaction can explain how the addition of sodium methanolate leads to FAME formation but does not explain that all methanol moieties introduced by the sodium methanolate end up as FAME. This is an important distinction since in practice, some water will be present in the oil and this will react with sodium methanolate to form sodium hydroxide and methanol; it is this free methanol that also ends up as FAME, for which we need the enolate mechanism as explanation.

When the enolate anion reacts with the free hydroxyl group in a partial glyceride, this leads to interesterification according to:

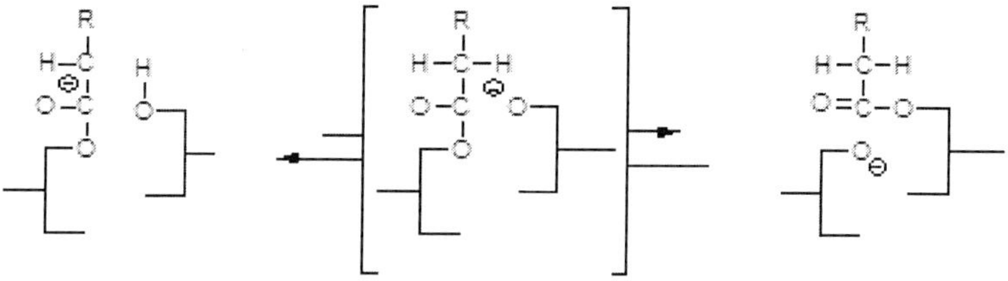

The fatty acid that was attached as an enolate to the glycerol moiety of the left ends up being attached to a glyceride on the right, which glyceride reacted with the enolate since it had a free hydroxyl group.

The above equation implies that the rate of interesterification of a glyceride oil should depend on its free hydroxyl (partial glyceride) content. This implication awaits experimental verification.

An essential aspect of this enolate mechanism is that it explains that on catalyst inactivation, an equivalent amount of free fatty acids or soap is formed according to:

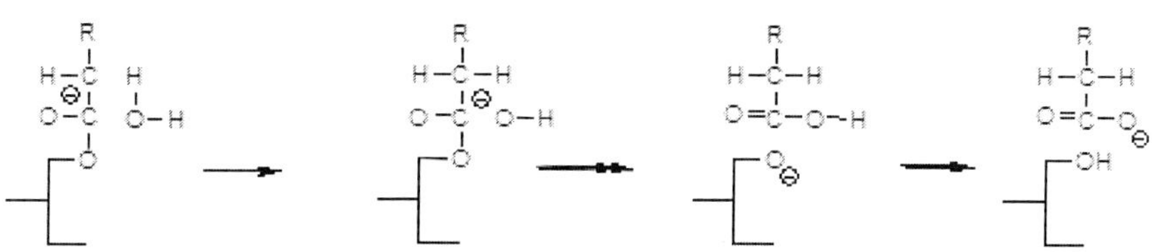

This soap formation could not be explained by the glycerolate mechanism. According to Rozendaal, the sodium hydroxide formed when the glycerolate reacts with water would saponify the glycerides present, but this reaction is slow and would not go to completion, whereas the soap formation has been observed to be quantitative. His suggestion that inactivating the catalyst with acid water would diminish the soap/FFA formation is not borne out by experiment either.

In several reaction equations above, the ester interchange reaction leads to the formation of a glycerolate anion. The enolate mechanism assumes this glycerolate anion to regenerate an enolate anion according to:

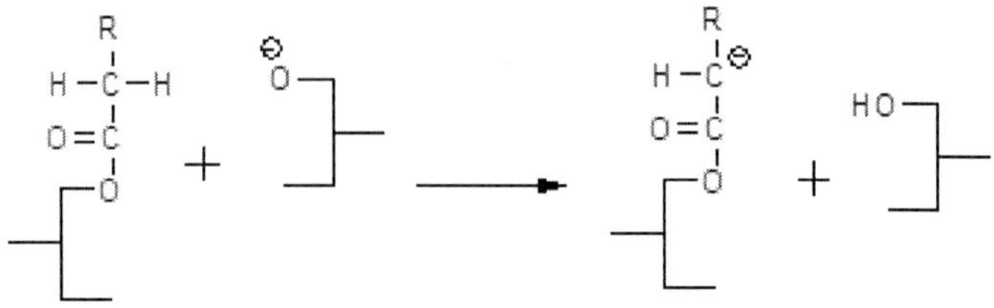

Another phenomenon that was not explained by the glycerolate mechanism is the fact that compounds like acetone accelerate the interesterification reaction. According to the enolate mechanism, a compound like acetone can react with the glycerolate anion according to:

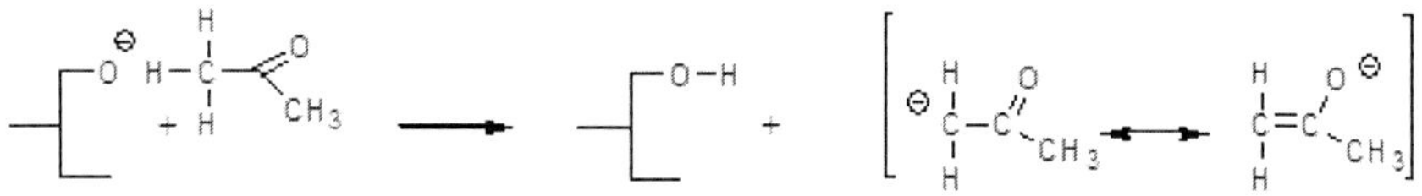

and form a highly mobile intermediate anion that subsequently reacts with a fatty acid moiety by abstracting an α-hydrogen and thereby regenerates the catalytic intermediate according to:

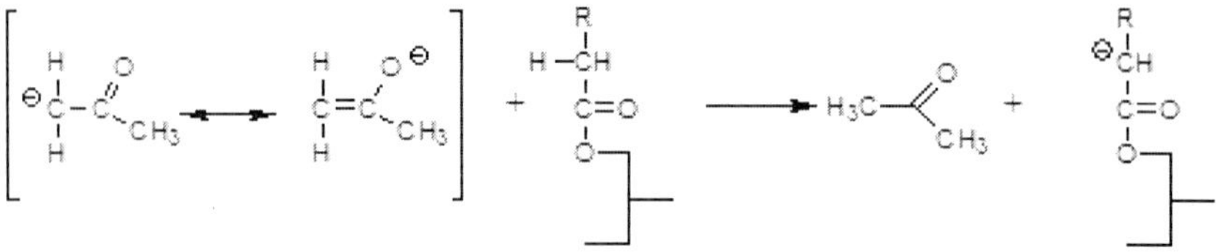

That the above reactions actually occur has been demonstrated experimentally by using triglycerides with fatty acid moieties the α-positions of which had been deuterated and interesterifying these triglycerides in the presence of acetone: in the course of the reaction, the acetone became deuterated.

In reactions where triglycerides react with other triglycerides (or rather partial glycerides), with alcohols such as methanol during the transesterification process to produce biodiesel from triglyceride oil, or with polyols such as glycerol that may be added as such (glycerolysis) or result as a by-product from other reactions (methanolysis), several equilibria between the various anions are involved. The positions of these equilibria are determined by the differences in thermodynamic properties of these anions and by the relative abundance of the reagents involved. The figure below shows the main equilibrium reactions involved:

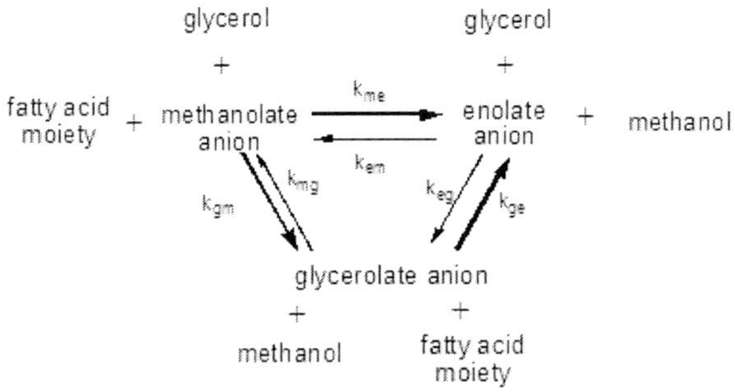

During a transesterification reaction in which triglycerides are allowed to react with methanol to from FAME and free glycerol, the hydroxyl concentration is large. Consequently, the enolate anion concentration will be much lower than during the interesterification of triglyceride oils when the free hydroxyl concentration is quite low.

The proposed enolate mechanism also provides an explanation for the observation that β-ketoesters are formed gradually and that the catalytic activity is thermally unstable. At low temperatures, the catalytic intermediate formed by the addition of sodium methanolate to triglyceride oil is active for a long time and allows a directed interesterification reaction to proceed for hours. Above 120°C, the intermediate starts being very active but it loses its activity in a matter of minutes. The mechanism proposes that the enolate anions are gradually converted into β-ketoesters and that the latter have hardly any catalytic activity.

Industrial Interesterification Process

On an industrial scale, the chemical interesterification process is most often carried out as a batch process. This has the advantage that different products can be made in the same installation without cross-contamination and this has the advantage that their inventories can be kept quite low. Another advantage of the interesterification process – and this applies to both chemical and enzymatic interesterification - is that it allows of a wide choice of the raw materials to be used. The process aims at producing a hardstock with certain physical properties. These determine the triglyceride composition of the product in broad terms and therefore also its fatty acid composition. I say 'broad terms' since replacing a linolenic acid moiety in a triglyceride by an linoleic acid moiety hardly affects the physical properties of a hardstock that really matter. Accordingly, the fat blend to be interesterified can be selected on the basis of fatty acid composition only, whereby the agricultural origin and the physical properties of the blend components are irrelevant.

So if palm kernel olein, a by-product of the fractionation of palm kernel oil, is the cheapest source of medium chain fatty acids and hydrogenated palm stearin the cheapest source of long chain saturated fatty acids, these can be profitably used as components of an interesterification fat blend, where only fatty acid composition matters.

Because sodium methanolate reacts with water and fatty acids, the fat blend to be interesterified must be dry and neutral. In practice, this is achieved by using neutralised components and ensuring their neutrality by adding some caustic soda to the interesterification reaction vessel and evaporating any moisture present by applying a vacuum. There is no need to use fully refined oil (RBD, refined, bleached, deodorised). During interesterification, some colour may be formed so that the interesterified product must be bleached and deodorisation is preferably carried out just prior to the product being used in or shipped to downstream manufacturing. Using RBD oil as raw material is therefore unnecessarily expensive.

When the oil blend to be interesterified has been dried, the interesterification catalyst can be added. If this is sodium methanolate, an amount of 0.05 wt% suffices. The reaction is very fast but usually allowed to proceed for some 30 minutes. Then the catalyst is inactivated by the addition of water, which leads to the formation of soaps. If water containing some acid is used for the inactivation, free fatty acids (FFA) are formed instead of soaps.

The soapy water can be removed by decantation and residual soaps can be removed by washing with water, treatment with a silica hydrogel or by adding some acid. This acid converts the soaps into FFA that are subsequently removed during vacuum steam stripping (deodorisation). This deodorisation process also ensures the removal of the fatty acid methyl esters (FAME) formed from the methanol introduced by the sodium methanolate catalyst.

Both the soaps/FFA and the FAME constitute a yield loss and this loss is considered to be the main disadvantage of the chemical interesterification process. Given the relative molecular mass of sodium methanolate of 54 and the stoichiometric production of soap/FFA and FAME, using 0.05 wt% of catalyst leads to a yield loss of at least 0.5

wt% oil. Using more catalyst would make this yield loss even higher. Trying to avoid or reduce this yield loss is therefore a worthwhile objective.

Avoiding the formation of FAME implies using a catalyst other than sodium methanolate or evaporating the methanol as soon as it has been formed. Since the latter possibility requires special equipment, it is therefore hardly opportune. Using another catalyst is certainly possible. The use of sodium glycerolate was disclosed in 1965 and subsequently it has also been used in continuous interesterification plants in Europe.

Using sodium or potassium metal or their alloy is another way to avoid introducing methanol into the interesterification reaction mixture that has been tested industrially. Whether methanol-free catalysts like sodium hydride and sodium amide have ever been used on an industrial scale is not clear from the literature.

Just as FAME formation can be avoided by preventing methanol from entering the reaction mixture, soap formation can be avoided by inactivating the catalyst in the absence of water. Since the next step involves treating the reaction mixture with bleaching earth, it would be quite elegant if acid activated bleaching earth were to be an effective reagent for catalyst inactivation, but this still has to be tried in a laboratory.

Enzymatic Catalysis

Enzyme catalysis is the increase in the rate of a chemical reaction by the active site of a protein. The protein catalyst (enzyme) may be part of a multi-subunit complex, and/or may transiently or permanently associate with a Cofactor (e.g. adenosine triphosphate). Catalysis of biochemical reactions in the cell is vital due to the very low reaction rates of the uncatalysed reactions. A key driver of protein evolution is the optimization of such catalytic activities via protein dynamics.

The mechanism of enzyme catalysis is similar in principle to other types of chemical catalysis. By providing an alternative reaction route the enzyme reduces the energy required to reach the highest energy transition state of the reaction. The reduction of activation energy (Ea) increases the amount of reactant molecules that achieve a sufficient level of energy, such that they reach the activation energy and form the product. As with other catalysts, the enzyme is not consumed during the reaction (as a substrate is) but is recycled such that a single enzyme performs many rounds of catalysis.

Induced Fit

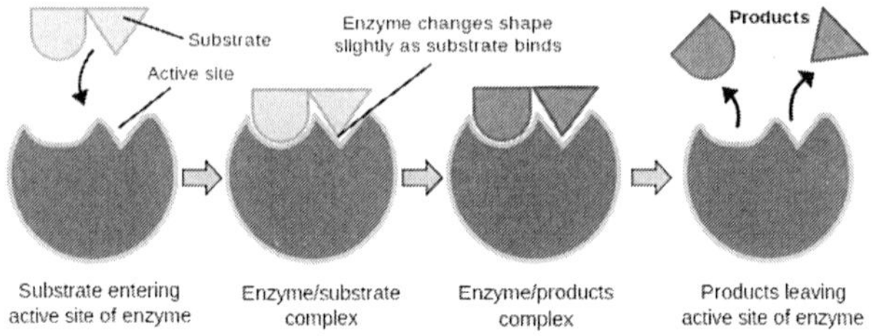

Figure: *Diagrams to show the induced fit hypothesis of enzyme action*

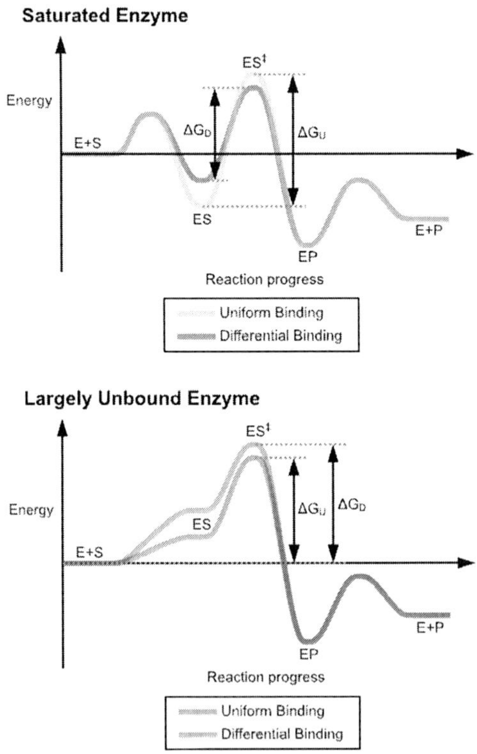

Figure: *The different mechanisms of substrate binding*

The favoured model for the enzyme-substrate interaction is the induced fit model. This model proposes that the initial interaction between enzyme and substrate is relatively weak, but that these weak interactions rapidly induce conformational changes in the enzyme that strengthen binding.

The advantages of the induced fit mechanism arise due to the stabilizing effect of strong enzyme binding. There are two different mechanisms of substrate binding: uniform binding, which has strong substrate binding, and differential binding, which has strong transition state binding. The stabilizing effect of uniform binding increases both substrate and transition state binding affinity, while differential binding increases only transition state binding affinity.

Both are used by enzymes and have been evolutionarily chosen to minimize the Ea of the reaction. Enzymes that are saturated, that is, have a high affinity substrate binding, require differential binding to reduce the Ea, whereas small substrate unbound enzymes may use either differential or uniform binding.

These effects have led to most proteins using the differential binding mechanism to reduce the Ea, so most proteins have high affinity of the enzyme to the transition state. Differential binding is carried out by the induced fit mechanism - the substrate first binds weakly, then the enzyme changes conformation increasing the affinity to the transition state and stabilizing it, so reducing the activation energy to reach it.

It is important to clarify, however, that the induced fit concept cannot be used to rationalize catalysis. That is, the chemical catalysis is defined as the reduction of $Ea^{\ddagger}$ (when the system is already in the $ES^{\ddagger}$) relative to $Ea^{\ddagger}$ in the uncatalyzed reaction in water (without the enzyme). The induced fit only suggests that the barrier is lower in the closed form of the enzyme but does not tell us what the reason for the barrier reduction is.

Induced fit may be beneficial to the fidelity of molecular recognition in the presence of competition and noise via the conformational proofreading mechanism .

Mechanisms of an Alternative Reaction Route

These conformational changes also bring catalytic residues in the active site close to the chemical bonds in the substrate that will be altered in the reaction. After binding takes place, one or more mechanisms of catalysis lowers the energy of the reaction's transition state, by providing an alternative chemical pathway for the reaction. There are six possible mechanisms of "over the barrier" catalysis as well as a "through the barrier" mechanism:

Bond Strain

This is the principal effect of induced fit binding, where the affinity of the enzyme to the transition state is greater than to the substrate itself. This induces structural rearrangements which strain substrate bonds into a position closer to the conformation of the transition state, so lowering the energy difference between the substrate and transition state and helping catalyze the reaction.

However, the strain effect is, in fact, a ground state destabilization effect, rather than transition state stabilization effect. Furthermore, enzymes are very flexible and they cannot apply large strain effect.

In addition to bond strain in the substrate, bond strain may also be induced within the enzyme itself to activate residues in the active site.

Proximity and Orientation

This increases the rate of the reaction as enzyme-substrate interactions align reactive chemical groups and hold them close together. This reduces the entropy of the reactants and thus makes reactions such as ligations or addition reactions more favourable, there is a reduction in the overall loss of entropy when two reactants become a single product.

This effect is analogous to an effective increase in concentration of the reagents. The binding of the reagents to the enzyme gives the reaction intramolecular character, which gives a massive rate increase.

For Example: Similar reactions will occur far faster if the reaction is intramolecular.

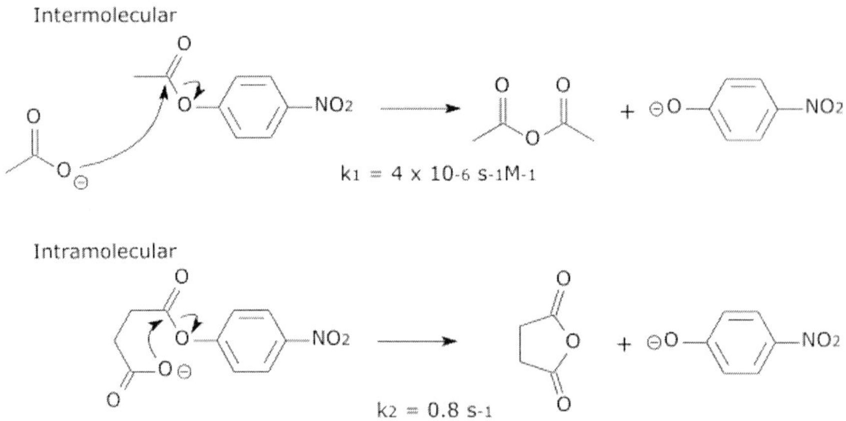

Figure: *The effective concentration of acetate in the intramolecular reaction can be estimated as* $k_2/k_1 = 2 \times 10^5$ *Molar.*

However, the situation might be more complex, since modern computational studies have established that traditional examples of proximity effects cannot be related directly to enzyme entropic effects. Also, the original entropic proposal has been found to largely overestimate the contribution of orientation entropy to catalysis.

Proton Donors or Acceptors

Proton donors and acceptors, i.e. acids and base may donate and accept protons in order to stabilize developing charges in the transition state.This typically has the effect of activating nucleophile and electrophile groups, or stabilizing leaving groups. Histidine is often the residue involved in these acid/base reactions, since it has a pKa close to neutral pH and can therefore both accept and donate protons.

Many reaction mechanisms involving acid/base catalysis assume a substantially altered pKa. This alteration of pKa is possible through the local environment of the residue.

Conditions	***Acids***	***Bases***
Hydrophobic environment	Increase pKa	Decrease pKa
Adjacent residues of like charge	Increase pKa	Decrease pKa
Salt bridge (and hydrogenbond) formation	Decrease pKa	Increase pKa

pKa can also be influenced significantly by the surrounding environment, to the extent that residues which are basic in solution may act as proton donors, and vice versa.

For example: Catalytic triad of a serine protease

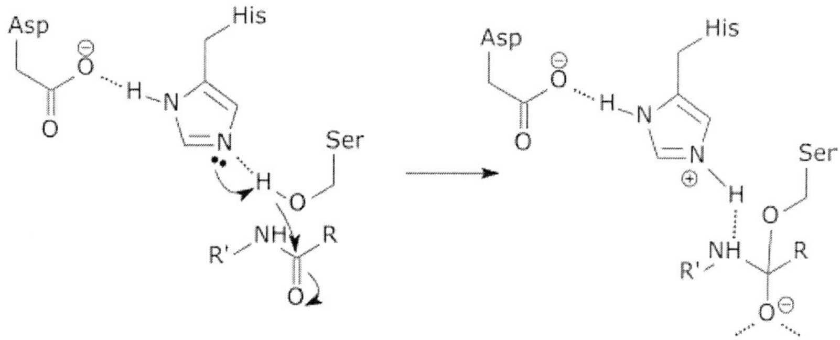

Figure: *The initial step of the serine protease catalytic mechanism involves the histidine of the active site accepting a proton from the serine residue. This prepares the serine as a nucleophile to attack the amide bond of the substrate. This mechanism includes donation of a proton from serine (a base, pKa 14) to histidine (an acid, pKa 6), made possible due to the local environment of the bases.*

It is important to clarify that the modification of the pKa's is a pure part of the electrostatic mechanism. Furthermore, the catalytic effect of the above example is mainly associated with the reduction of the pKa of the oxyanion and the increase in the pKa of the histidine, while the proton transfer from the serine to the histidine is not catalyzed significantly, since it is not the rate determining barrier.

Electrostatic Catalysis

Stabilization of charged transition states can also be by residues in the active site forming ionic bonds (or partial ionic charge interactions) with the intermediate. These bonds can either come from acidic or basic side chains found on amino acids such as lysine, arginine, aspartic acid or glutamic acid or come from metal cofactors such as zinc. Metal ions are particularly effective and can reduce the pKa of water enough to make it an effective nucleophile.

Systematic computer simulation studies established that electrostatic effects give, by far, the largest contribution to catalysis. In particular, it has been found that enzyme provides an environment which is more polar than water, and that the ionic transition states are stabilized by fixed dipoles. This is very different from transition state stabilization in water, where the water molecules must pay with "reorganization energy". In order to stabilize ionic and charged states. Thus, the catalysis is associated with the fact that the enzyme polar groups are preorganized

The magnitude of the electrostatic field exerted by an enzyme's active site has been shown to be highly correlated with the enzyme's catalytic rate enhancement

For Example: Carboxypeptidase catalytic mechanism

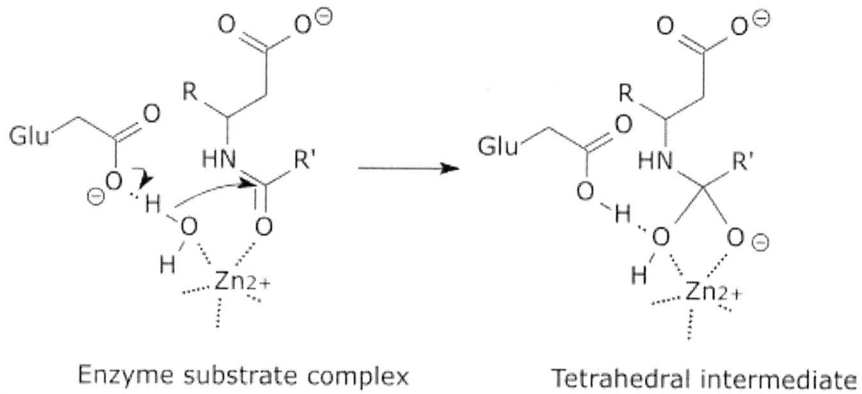

Figure: *The tetrahedral intermediate is stabilised by a partial ionic bond between the Zn^{2+} ion and the negative charge on the oxygen.*

Binding of substrate usually excludes water from the active site, thereby lowering the local dielectric constant to that of an organic solvent. This strengthens the electrostatic interactions between the charged/polar substrates and the active sites. In addition, studies have shown that the charge distributions about the active sites are arranged so as to stabilize the transition states of the catalyzed reactions. In several enzymes, these charge distributions apparently serve to guide polar substrates towards their binding sites so that the rates of these enzymatic reactions are greater than their apparent diffusion-controlled limits.

Covalent Catalysis

Covalent catalysis involves the substrate forming a transient covalent bond with residues in the active site or with a cofactor. This adds an additional covalent intermediate to the reaction, and helps to reduce the energy of later transition states of the reaction. The covalent bond must, at a later stage in the reaction, be broken to regenerate the enzyme. This mechanism is utilised by the catalytic triad of enzymes such as proteases like chymotrypsin and trypsin, where an acyl-enzyme intermediate is formed. Schiff base formation using the free amine from a lysine residue is another mechanism, as seen in the enzyme aldolase during glycolysis.

Some enzymes utilize non-amino acid cofactors such as pyridoxal phosphate (PLP) or thiamine pyrophosphate (TPP) to form covalent intermediates with reactant molecules. Such covalent intermediates function to reduce the energy of later transition states, similar to how covalent intermediates formed with active site amino acid residues allow stabilization, but the capabilities of cofactors allow enzymes to carryout reactions that amino acid side residues alone could not. Enzymes utilizing such cofactors include the PLP-dependent enzyme aspartate transaminase and the TPP-dependent enzyme pyruvate dehydrogenase.

It is important to clarify that covalent catalysis does correspond in most cases to simply the use of a specific mechanism rather than to true catalysis. For example, the energetics of the covalent bond to the serine molecule in chymotrypsin should be compared to the well-understood covalent bond to the nucleophile in the uncatalyzed solution reaction. A true proposal of a covalent catalysis (where the barrier is lower than the corresponding barrier in solution) would require, for example, a partial covalent bond to the transition state by an enzyme group (e.g., a very strong hydrogen bond), and such effects do not contribute significantly to catalysis.

Quantum Tunneling

These traditional "over the barrier" mechanisms have been challenged in some cases by models and observations of "through the barrier" mechanisms (quantum tunneling). Some enzymes operate with kinetics which are faster than what would be predicted by the classical $\Delta G^{\ddagger}$. In "through the barrier" models, a proton or an electron can tunnel through activation barriers. Quantum tunneling for protons has been observed in tryptamine oxidation by aromatic amine dehydrogenase.

Interestingly, quantum tunneling does not appear to provide a major catalytic advantage, since the tunneling contributions are similar in the catalyzed and the uncatalyzed reactions in solution. However, the tunneling contribution (typically enhancing rate constants by a factor of ~1000 compared to the rate of reaction for the classical 'over the barrier' route) is likely crucial to the viability of biological organisms. This emphasizes the general importance of tunneling reactions in biology.

In 1971-1972 the first quantum-mechanical model of enzyme catalysis was formulated.

Active Enzyme

The binding energy of the enzyme-substrate complex cannot be considered as an external energy which is necessary for the substrate activation. The enzyme of high energy content may firstly transfer some specific energetic group X_1 from catalytic site of the enzyme to the final place of the first bound reactant, then another group X_2 from the second bound reactant (or from the second group of the single reactant) must be transferred to active site to finish substrate conversion to product and enzyme regeneration.

We can present the whole enzymatic reaction as a two coupling reactions:

$$S_1 + EX_1 \Rightarrow S_1EX_1 \Rightarrow P_1 + EP_2 \quad (1)$$

$$S_2 + EP_2 \Rightarrow S_2EP_2 \Rightarrow P_2 + EX_2 \quad (2)$$

It may be seen from reaction (1) that the group X_1 of the active enzyme appears in the product due to possibility of the exchange reaction inside enzyme to avoid both electrostatic inhibition and repulsion of atoms. So we represent the active enzyme as a powerful reactant of the enzymatic reaction. The reaction (2) shows incomplete conversion of the substrate because its group X_2 remains inside enzyme. This approach as idea had formerly proposed relying on the hypothetical extremely high enzymatic conversions (catalytically perfect enzyme).

The crucial point for the veriûcation of the present approach is that the catalyst must be a complex of the enzyme with the transfer group of the reaction. This chemical aspect is supported by the well-studied mechanisms of the several enzymatic reactions. Let us consider the reaction of peptide bond hydrolysis catalyzed by a pure protein α-chymotrypsin (an enzyme acting without a cofactor), which is a well-studied member of the serine proteases family.

We present the experimental results for this reaction as two chemical steps:

$$S_1 + EH \Rightarrow P_1 + EP_2 \quad (3)$$

$$EP_2 + H\text{–}O\text{–}H \Rightarrow EH + P_2 \quad (4)$$

where S_1 is a polypeptide, P_1 and P_2 are products. The ûrst chemical step (3) includes the formation of a covalent acyl-enzyme intermediate. The second step (4) is the deacylation step. It is important to note that the group H+, initially found on the enzyme, but not in water, appears in the product before the step of hydrolysis, therefore it may be considered as an additional group of the enzymatic reaction.

Thus, the reaction (3) shows that the enzyme acts as a powerful reactant of the reaction. According to the proposed concept, the H transport from the enzyme promotes the ûrst reactant conversion, breakdown of the ûrst initial chemical bond (between groups P_1 and P_2). The step of hydrolysis leads to a breakdown of the second chemical bond and regeneration of the enzyme.

The proposed chemical mechanism does not depend on the concentration of the substrates or products in the medium. However, a shift in their concentration mainly causes free energy changes in the ûrst and ûnal steps of the reactions (1) and (2) due to the changes in the free energy content of every molecule, whether S or P, in water solution. This approach is in accordance with the following mechanism of muscle contraction. The ûnal step of ATP hydrolysis in skeletal muscle is the product release caused by the association of myosin heads with actin. The closing of the actin-binding cleft during the association reaction is structurally coupled with the opening of the nucleotide-binding pocket on the myosin active site.

Notably, the ûnal steps of ATP hydrolysis include the fast release of phosphate and the slow release of ADP. The release of a phosphate anion from bound ADP anion into water solution may be considered as an exergonic reaction because the phosphate anion has low molecular mass.

Thus, we arrive at the conclusion that the primary release of the inorganic phosphate $H_2PO_4^-$ leads to transformation of a signiûcant part of the free energy of ATP hydrolysis into the kinetic energy of the solvated phosphate, producing active streaming. This assumption of a local mechano-chemical transduction is in accord with Tirosh's mechanism of muscle contraction, where the muscle force derives from an integrated action of active streaming created by ATP hydrolysis.

Examples of Catalytic Mechanisms

In reality, most enzyme mechanisms involve a combination of several different types of catalysis.

Triose Phosphate Isomerase

Triose phosphate isomerase (EC 5.3.1.1) catalyses the reversible interconvertion of the two triose phosphates isomers dihydroxyacetone phosphate and D-glyceraldehyde 3-phosphate.

Trypsin

Trypsin (EC 3.4.21.4) is a serine protease that cleaves protein substrates at lysine and arginine amino acid residues using a catalytic triad of active site residues to perform nucleophilic, covalent catalysis.

Aldolase

Aldolase (EC 4.1.2.13) catalyses the breakdown of fructose 1,6-bisphosphate (F-1,6-BP) into glyceraldehyde 3-phosphate and dihydroxyacetone phosphate (DHAP).

Enzyme Diffusivity

The advent of single-molecule studies led in the 2010s to the observation that the movement of untethered enzymes increases with increasing substrate concentration and increasing reaction enthalpy. Subsequent observations suggest that this increase in diffusivity is driven by transient displacement of the enzyme's centre of mass, resulting in a "recoil effect that propels the enzyme".

Reaction Similarity

Similarity between enzymatic reactions (EC) can be calculated by using bond changes, reaction centres or substructure metrics (EC-BLAST).

Fractionation

Fractionation is a separation process in which a certain quantity of a mixture (gas, solid, liquid, suspension or isotope) is divided during a phase transition, into a number of smaller quantities (fractions) in which the composition varies according to a gradient. Fractions are collected based on differences in a specific property of the individual components. A common trait in fractionations is the need to find an optimum between the amount of fractions collected and the desired purity in each fraction. Fractionation makes it possible to isolate more than two components in a mixture in a single run. This property sets it apart from other separation techniques.

Fractionation is widely employed in many branches of science and technology. Mixtures of liquids and gases are separated by fractional distillation by difference in boiling point. Fractionation of components also takes place in column chromatography by a difference in affinity between stationary phase and the mobile phase. In fractional crystallization and fractional freezing, chemical substances are fractionated based on difference in solubility at a given temperature. In cell fractionation, cell components are separated by difference in mass.

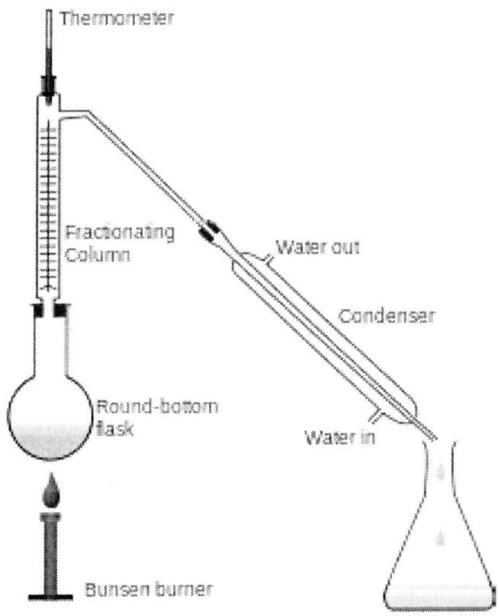

Figure: *Fractional distillation apparatus using a Liebig condenser. A conical flask is used as a receiving flask. Here the distillation head and fractionating column are combined in one piece.*

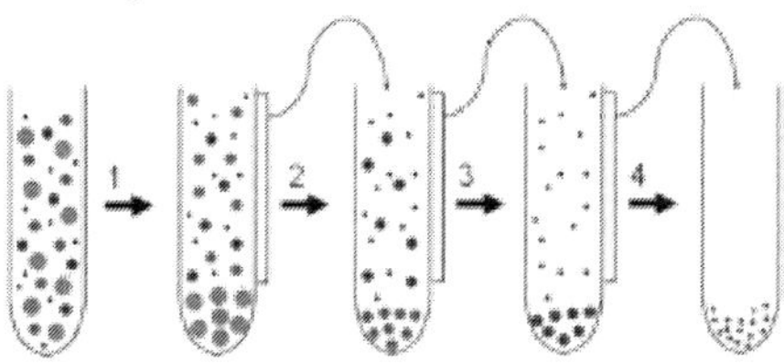

Figure: *Differential centrifugation.*

Fractionation Products

Fats and oils are fractionated in order to enhance their working value and/or to increase their commercial value, or to make special products. Fractionation can be applied to crude oils, to deacidified and bleached oils or to oils that have only been bleached. Fats and oils are not homogeneous substances, but mixtures of various triglycerides with differing melting points. This affects their physical properties. Thus, fats and oils do not have a definite melting point, but a melting range. This characteristic is exploited for separation purposes. Fractionation is a purely physical process, in which both fat and triglyceride components remain unchanged. Separation is governed exclusively by the melting point or, in the case of wet fractionation, by solubility in a suitable solvent. The liquid components of a fat are present as droplets in the crystal lattice of components with higher melting-points and are squeezed out. Despite the use of modern equipment, the process takes several hours and can be broken down into the following stages:

- Melting or dissolving the fat/oil
- Conditioning
- Crystallising
- Separating

The three methods currently used for fractionation are:

Dry Fractionation (Without Adjuvants)

Dry fractionation is based upon the principle of separation according to the differing melting points for the various components, e.g. olein and stearin. The process has the advantage of being purely physical, involving no additives of any kind and, thus, dispensing with the need for subsequent product purification. One disadvantage is relatively poor separation, compared with the two other processes. Dry fractionation is performed by heating the mixture to a temperature above the melting point and then leaving it to cool and separate, usually into two fractions.

Lanza Fractionation

The principle is based on a patent granted to the Lanza brothers in 1905. Separation is not performed mechanically, but by wetting the crystal surface with aqueous detergents, into which salts may also be mixed .

This hinders emulsification. Wetting causes higher melting-point components in the oil to become hydrophilic and to migrate into the aqueous phase. This produces large crystal-free oil droplets that then coalesce. In the next step in processing, the two phases are separated by centrifugation.

Wet Fractionation

The principle is based upon differences in solvent solubility at a given temperature. This results in sharp separation of the different phases, which can be controlled by adjusting the temperature and the amount of solvent. If particularly pure fractions are required, the product can be re-washed with a solvent and the latter is subsequently removed.

Winterising

Winterising is a very simple form of fractionation, in which it is not two fractions with similar particle sizes that are separated, but small portions that are separated from the bulk of the oil. This process is not used to modify an oil, but to alter its behaviour, so that it becomes easier to handle at low temperatures. At refrigerator temperatures, waxes, along with high melting-point glycerides, tend to curdle. This is sometimes mistaken for spoiling.

De-acidified and bleached oil is cooled down to 5-15°C. Unwanted waxes, mucilages and high melting-point glycerides precipitate out and are filtered off. In most cases, this filtering process is problematic, due either to the small amounts of waxes and glycerides involved or to the properties of the substances that need to be separated. This problem can be tackled by adding small amounts of fuller's earth.

Winterising can be performed in batches or continuously and takes somewhere in the region of 4-8 hours, depending on the type of oil and on the process used (batch or continuous).

There is also the option of solvent winterising. The oil is mixed with the solvent (to avoid losses) and cooled. The high melting-point crystals in the triglycerides are filtered off and the solvent is then removed.

Chapter 5

Fats For Chocolate and Sugar Confectionery

Confectionery is the art of making confections, which are food items that are rich in sugar and carbohydrates. Exact definitions are difficult. In general, though, confectionery is divided into two broad and somewhat overlapping categories, bakers' confections and sugar confections.

Bakers' confectionery, also called flour confections, includes principally sweet pastries, cakes, and similar baked goods. In the Middle East and Asia, flour-based confections are more dominant.

Sugar confectionery includes sweets, candied nuts, chocolates, chewing gum, sweetmeats, pastillage, and other confections that are made primarily of sugar. In some cases, chocolate confections (confections made of chocolate) are treated as a separate category, as are sugar-free versions of sugar confections. The words candy (US and Canada), sweets (UK and Ireland), and lollies (Australia and New Zealand) are common words for the most common varieties of sugar confectionery.

The confectionery industry also includes specialized training schools and extensive historical records. Traditional confectionery goes back to ancient times, and continued to be eaten through the Middle Ages into the modern era.

Generally, confections are low in micronutrients and protein but high in calories. They may be fat-free foods, although some confections, especially fried doughs, are high-fat foods. Many confections are considered empty calories. Specially formulated chocolate has been manufactured in the past for military use as a high-density food energy source.

Sweetening Agents

Confections are defined by the presence of sweeteners. These are usually sugars, but it is possible to buy sugar-free sweets, such as sugar-free peppermints. The most common sweetener for home cooking is table sugar, which is chemically a disaccharide called sucrose. Hydrolysis of sucrose gives a mixture called invert sugar, which is sweeter and is also a common commercial ingredient. Finally confections, especially commercial ones, are sweetened by a variety of syrups obtained by hydrolysis of starch. These sweeteners include all types of corn syrup.

Bakers' Confectionery

Figure: *Petits fours are baker's confections.*

Bakers' confectionery includes sweet baked goods, especially those that are served for the dessert course. Bakers' confections are sweet foods that feature flour as a main ingredient and are baked. Major categories include cakes, sweet pastries, doughnuts, scones, and cookies.

Sugar Confectionery

Sugar confections include sweet, sugar-based foods, which are usually eaten as snack food. This includes sugar candies, chocolates, candied fruits and nuts, chewing gum, and sometimes ice cream.

The United Nations' International Standard Industrial Classification of All Economic Activities (ISIC) scheme (revision 4) classifies both chocolate and sugar confectionery as ISIC 1073, which includes the manufacture of chocolate and chocolate confectionery; sugar confectionery proper (caramels, cachous, nougats, fondant, white chocolate), chewing gum, preserving fruit, nuts, fruit peels, and making confectionery lozenges and pastilles.

In the European Union, the Statistical Classification of Economic Activities in the European Community (NACE) scheme (revision 2) matches the UN classification, under code number 10.82. In the United States, the North American Industry Classification System (NAICS 2012) splits sugar confectionery across three categories: National industry code 311340 for all non-chocolate confectionery manufacturing, 311351 for chocolate and confectionery manufacturing from cacao beans, and national industry 311352 for confectionery manufacturing from purchased chocolate. Ice cream and sorbet are classified with dairy products under ISIC 1050, NACE 10.52, and NAICS 311520.

Different Dialects of English use Regional Terms for Sugar Confections

- In Britain, Ireland and some Commonwealth countries, *sweets* or, more colloquially, *sweeties* (particularly used by children, the Scottish Gaelic word *suiteis* is a derivative).
- In Australia and New Zealand, *lollies*. *Chuddy* is an Australian slang for chewing gum.
- In North America, *candy*, although this term generally refers to a specific range of confectionery and does not include some items of sugar confectionery (e.g. ice cream). *Sweet* is occasionally used, as well as *treat*.

Examples

Sugar confectionery items include sweets, lollipops, candy bars, chocolate, cotton candy, and other sweet items of snack food. Some of the categories and types of sugar confectionery include the following:

- Caramels: Derived from a mixture of sucrose, glucose syrup, and milk products. The mixture does not crystallize, thus remains tacky.
- Chocolates: Bite-sized confectioneries generally made with chocolate.

- Divinity: A nougat-like confectionery based on egg whites with chopped nuts.
- Dodol: A toffee-like food delicacy popular in Indonesia, Malaysia, and the Philippines
- Dragée: Sugar-coated almonds and other types of sugar panned candy.
- Fondant: Prepared from a warm mixture of glucose syrup and sucrose, which is partially crystallized. The fineness of the crystallites results in a creamy texture.
- Fudge: Made by boiling milk and sugar to the soft-ball stage. In the US, it tends to be chocolate-flavoured.
- Halvah: Confectionery based on tahini, a paste made from ground sesame seeds.
- Hard candy: Based on sugars cooked to the hard-crack stage. Examples include suckers (known as *boiled sweets* in British English), lollipops, jawbreakers (or gobstoppers), lemon drops, peppermint drops and disks, candy canes, rock candy, etc. Also included are types often mixed with nuts such as brittle. Others contain flavourings including coffee such as Kopiko.
- Ice cream: Frozen, flavoured cream, often containing small pieces of chocolate, fruits and/or nuts.
- Jelly candies: Including those based on sugar and starch, pectin, gum, or gelatin such as Turkish delight (lokum), jelly beans, gumdrops, jujubes, gummies, etc.
- Liquorice: Containing extract of the liquorice root. Chewier and more resilient than gum/gelatin candies, but still designed for swallowing. For example, Liquorice allsorts. Has a similar taste to star anise.
- Marshmallow: "Peeps" (a trade name), circus peanuts, fluffy puff, Jet-Puffed Marshmallows etc.
- Marzipan: An almond-based confection, doughy in consistency, served in several different ways.
- Mithai: A generic term for confectionery in India, typically made from dairy products and/or some form of flour. Sugar or molases are used as sweeteners.
- Tablet: A crumbly milk-based soft and hard candy, based on sugars cooked to the soft-ball stage. Comes in several forms, such as wafers and heart shapes. Not to be confused with tableting, a method of candy production.

- Taffy or chews: A candy that is folded many times above 50 °C, incorporating air bubbles thus reducing its density and making it opaque.

Cultural Roles

Both bakers' and sugar confections are used to offer hospitality to guests.

Confections are used to mark celebrations or events, such as a wedding cake, birthday cake or Halloween.

Tourists commonly eat confections as part of their travels. The indulgence in rich, sugary foods is seen as a special treat, and choosing local specialties is popular.

For example, visitors to Vienna eat Sachertorte and visitors to seaside resorts in the UK eat Blackpool rock candy. Transportable confections like fudges and tablet may be purchased as souvenirs.

Production and Properties

The Production of Chocolate

Chocolate is a key ingredient in many foods such as milk shakes, candy bars, cookies and cereals. It is ranked as one of the most favourite flavours in North America and Europe. Despite its popularity, most people do not know the unique origins of this popular treat. Chocolate is a product that requires complex procedures to produce.

The process involves harvesting coca, refining coca to cocoa beans, and shipping the cocoa beans to the manufacturing factory for cleaning, coaching and grinding.

These cocoa beans will then be imported or exported to other countries and be transformed into different type of chocolate products.

Harvesting Cocoa & Cocoa Processing

Chocolate production starts with harvesting coca in a forest. Cocoa comes from tropical evergreen Cocoa trees, such as Theobroma Cocoa, which grow in the wet lowland tropics of Central and South America, West Africa and Southeast Asia (within 20 C of the equator) (Walter,1981) . Cocoa needs to be harvested manually in the forest. The seed pods of coca will first be collected; the beans will be selected and placed in piles.

These cocoa beans will then be ready to be shipped to the manufacturer for mass production.

Step #1: Plucking and Opening the Pods

Cocoa beans grow in pods that sprout off of the trunk and branches of cocoa trees. The pods are about the size of a football. The pods start out green and turn orange when they're ripe. When the pods are ripe, harvesters travel through the cocoa orchards with machetes and hack the pods gently off of the trees.

Machines could damage the tree or the clusters of flowers and pods that grow from the trunk, so workers must be harvest the pods by hand, using short, hooked blades mounted on long poles to reach the highest fruit.

After the cocoa pods are collected into baskets ,the pods are taken to a processing house. Here they are split open and the cocoa beans are removed. Pods can contain upwards of 50 cocoa beans each. Fresh cocoa beans are not brown at all, they do not taste at all like the sweet chocolate they will eventually produce.

Step #2: Fermenting the Cocoa Seeds

Now the beans undergo the fermentation processing. They are either placed in large, shallow, heated trays or covered with large banana leaves. If the climate is right, they may be simply heated by the sun. Workers come along periodically and stir them up so that all of the beans come out equally fermented. During fermentation is when the beans turn brown. This process may take five or eight days.

Step #3: Drying the Cocoa Seeds

After fermentation, the cocoa seeds must be dried before they can be scooped into sacks and shipped to chocolate manufacturers. Farmers simply spread the fermented seeds on trays and leave them in the sun to dry. The drying process usually takes about a week and results in seeds that are about half of their original weight.

The Dried and Roasted Cocoa Beans

Manufacturing Chocolate

Once the cocoa beans have reached the machinery of chocolate factories, they are ready to be refined into chocolate. Generally, manufacturing processes differ slightly due to the different species of cocoa trees, but most factories use similar machines to break down the cocoa beans into cocoa butter and chocolate (International Cocoa Organization, 1998). Firstly, fermented and dried cocoa beans will be refined to a roasted nib by winnowing and roasting. Then, they will be heated and will melt into chocolate liquor. Lastly, manufacturers blend chocolate liquor with sugar and milk to add flavour. After the blending process, the liquid chocolate will be stored or delivered to the molding factory in tanks and will be poured into moulds for sale. Finally, wrapping and packaging machines will pack the chocolates and then they will be ready to transport.

A Diagram Showing the Manufacturing Process

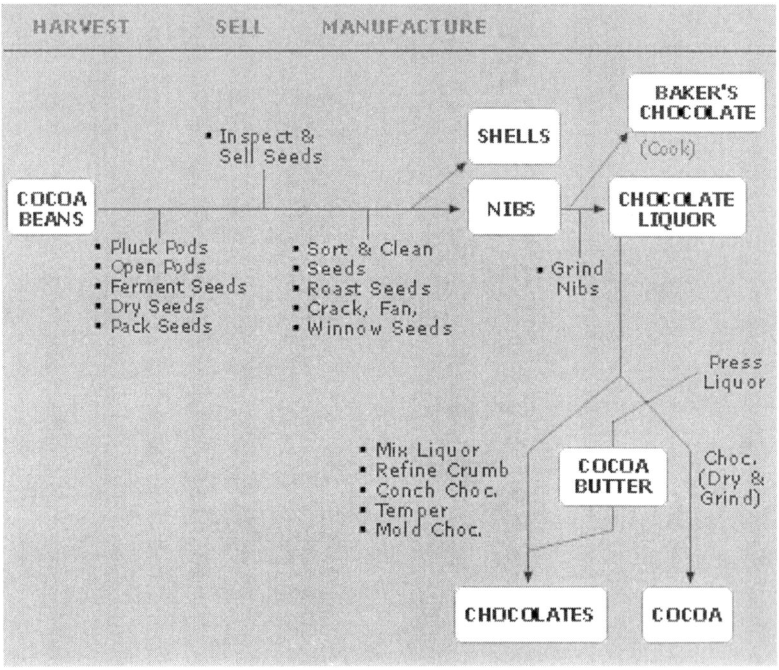

Step #1: Roasting and Winnowing the Cocoa

The first thing that chocolate manufacturers do with cocoa beans is roast them. This develops the colour and flavour of the beans into what our modern palates expect from fine chocolate. The outer shell of the beans is removed, and the inner cocoa bean meat is broken into small pieces called "cocoa nibs."

The roasting process makes the shells of the cocoa brittle, and cocoa nibs pass through a series of sieves, which strain and sort the nibs according to size in a process called "winnowing".

Step #2: Grinding the Cocoa Nibs

Grinding is the process by which cocoa nibs are ground into " cocoa liquor", which is also known as unsweetened chocolate or cocoa mass. The grinding process generates heat and the dry granular consistency of the cocoa nib is then turned into a liquid as the high amount of fat contained in the nib melts.

The cocoa liquor is mixed with cocoa butter and sugar. In the case of milk chocolate, fresh, sweetened condensed or roller-dry low-heat powdered whole milk is added, depending on the individual manufacturer's formula and manufacturing methods.

Step #3: Blending Cocoa Liquor and Molding Chocolate

After the mixing process, the blend is further refined to bring the particle size of the added milk and sugar down to the desired fineness. The Cocoa powder or 'mass' is blended back with the butter and liquor in varying quantities to make different types of chocolate or couverture. The basic blends with ingredients roughly in order of highest quantity first are as follows:

Milk Chocolate - sugar, milk or milk powder, cocoa powder, cocoa liquor, cocoa butter, Lethicin and Vanilla.

White Chocolate- sugar, milk or milk powder, cocoa liquor, cocoa butter, Lethicin and Vanilla.

Plain Dark Chocolate - cocoa powder, cocoa liquor, cocoa butter, sugar, Lethicin and Vanilla.

After blending is complete, molding is the final procedure for chocolate processing. This step allows cocoa liquor to cool and harden into different shapes depending on the mold. Finally the chocolate is packaged and distributed around the world.

Sugar Confectionery Production

Sugar confectionery refers to a large range of food items, commonly known as sweets. Boiled sweets, toffees, marshmallows, and fondant are all examples.

Sweets are a non-essential commodity, but are consumed by people from most income groups. The variety of products is enormous, ranging from cheap, individually-wrapped sweets, to those presented in boxes with sophisticated packaging.

Nutritional Significance

The main ingredient used in the production of sweets is sugar (sucrose). There is a danger that if sweets are consumed in excess over a prolonged period of time they may contribute to obesity. Unless good dental care is practiced, over-consumption can also lead to tooth decay.

Principles of Sugar Confectionery Production

By varying the ingredients used, the temperature of boiling, and the method of shaping, it is possible to make a wide variety of products. In all cases, however, the principle of production remains the same and is outlined below:

- balance the recipe
- prepare the ingredients
- mix together the ingredients
- boil the mixture until the desired temperature has been reached
- cool
- shape
- pack.

Many factors affect the production and storage of sweets:

- the degree of sucrose inversion
- the time and temperature of boiling
- the residual moisture content in the confectionery
- the addition of other ingredients.

Degree of Inversion

Sweets containing high concentrations of sugar (sucrose) may crystallize either during manufacture or on storage (commonly referred to as graining). Although this may be desirable for certain products (such as fondant and fudge), in most other cases it is seen as a quality defect.

When a sugar solution is heated, a certain percentage of sucrose breaks down to form 'invert sugar'. This invert sugar inhibits sucrose crystallization and increases the overall concentration of sugars in the mixture. This natural process of inversion, however, makes it difficult to accurately assess the degree of invert sugar that will be produced.

As a way of controlling the amount of inversion, certain ingredients, such as cream of tartar or citric acid, may be used. Such ingredients accelerate the breakdown of sucrose into invert sugar, and thereby increase the overall percentage of invert sugar in the solution. A more accurate method of ensuring the correct balance of invert sugar is to add glucose syrup, as this will directly increase the proportion of invert sugar in the mixture.

The amount of invert sugar in the sweet must be controlled, as too much may make the sweet prone to take up water from the air and become sticky. Too little will be insufficient to prevent crystallization of the sucrose. About 10-15 per cent of invert sugar is the amount required to give a non-crystalline product.

Time and Temperature of Boiling

The temperature of boiling is very important, as it directly affects the final sugar concentration and moisture content of the sweet. For a fixed concentration of sugar, a mixture will boil at the same temperature at the same altitude above sea-level, and therefore each type of sweet has a different heating temperature.

Boiling Point of Sucrose Solutions

Sucrose concentration (per cent)	*Degrees C Boiling point* *	*Degrees F Boiling point* *
40	101.4	214.5
50	102	215.5
60	103	217.5
70	105.5	222
75	108	227
80	111	232
85	116	241
90	122	252
95	130	266

*at sea level.

Variations in boiling temperature can make a difference between a sticky, cloudy sweet or a dry, clear sweet. An accurate way of measuring the temperature is to use a sugar thermometre. Other tests can be used to assess the temperature (for example, toffee temperatures can be estimated by removing a sample, cooling it in water, and examining it when cold). The temperatures are known by distinctive names such as 'soft ball', 'hard ball' etc., all of which refer to the consistency of the cold toffee.

Type of sweet	*Temperature range for boiling (Degrees C)*
Fondants	116-121
Fudge	116
Caramels and regular toffee	118-132
Hard toffee (e.g. butterscotch)	146-154
Hard-boiled sweets	149-166

Moisture Content

The water left in the sweet will influence its storage behaviour and determine whether the product will dry out, or pick up, moisture. For sweets which contain more than 4 per cent moisture, it is likely that sucrose will crystallize on storage. The surface of the sweet will absorb water, the sucrose solution will subsequently weaken, and crystallization will occur at the surface - later spreading throughout the sweet.

Added Ingredients

The addition of certain ingredients can affect the temperature of boiling. For example, if liquid milk is used in the production of toffees, the moisture content of the mixture immediately increases, and will therefore require a longer boiling time in order to reach the desired moisture content.

Added ingredients also have an effect on the shelf-life of the sweet. Toffees, caramels, and fudges, which contain milk-solids and fat, have a higher viscosity, which controls crystallization. On the other hand, the use of fats may make the sweet prone to rancidity, and consequently the shelf-life will be shortened.

Types of Sweets

Fondants and Creams

Fondant is made by boiling a sugar solution with the optional addition of glucose syrup. The mixture is boiled to a temperature in the range of 116-121°C, cooled, and then beaten in order to control the crystallization process and reduce the size of the crystals.

Creams are fondants which have been diluted with a weak sugar solution or water. These products are not very stable due to their high water content, and therefore have a shorter shelf-life than many other sugar confectionery products. Both fondants and creams are commonly used as soft centres for chocolates and other sweets.

Gelatin Sweets

These sweets include gums, jellies, pastilles, and marshmallows. They are distinct from other sweets as they have a rather spongy texture which is set by gelatin.

Toffee and Caramels

These are made from sugar solutions with the addition of ingredients such as milk-solids and fats. Toffees have a lower moisture content than caramels and consequently have a harder texture. As the product does not need to be clear, it is possible to use unrefined sugar such as jaggery or gur, instead of white granular sugar.

Hard-boiled Sweets

These are made from a concentrated solution of sugar which has been heated and then cooled to form a solid mass containing less than 2 per cent moisture. Within this group of products there is a wide scope to create many different colours, flavours and shapes through the use of added flavourings and colourings.

Table: *The table below outlines the processing stages for a selected range of confectionery items.*

	Mix ingredients	*Boil*	*Cool*	*Beat*	*Form/set*
Hard-boiled sweets	*	*	*		*
Fondant	*	*	*	*	*
Toffees/caramels	*	*	*		*
Fudge	*	*	*	*	*
Jellies	*	*	*		*
Marshmallows	*	*	*	*	*

Processing

Boiling

There are three main ways by which to boil the sugar solution:

- a simple open boiling pan
- a steam jacketed pan
- a vacuum cooker.

Steam jacketed pans are often fitted with scrapers and blades which make the mixing and heating process more uniform, and lessen the possibility of localized over-heating. Vacuum cookers are not generally used at a small scale.

Cooling

All sweets are cooled slightly before being shaped. Most simply, the boiled mass is poured onto a table (this should be made from metal, stone, or marble to cool the product uniformly). The table should be clean and free from cracks, as they may harbour dirt and microorganisms. It is important that the boiled mass is cooled sufficiently, since if it is to be formed by hand there is a danger that the operator may suffer burns.

Equipment Required

Processing stage	*Equipment*	*Section reference*
Mix ingredients	Weighing and measuring equipment	64.1 and 64.2
Boil	Heat source	36.0
	Boiling pans	48.0
	Steam jacketed pans	48.0
	Thermometer	63.0
Cool	Table	
Beat	Hand whisk or liquid mixer	43.1
Form/set	Starch mould cutting equipment	17.1
Pack	Waxed papers cellulose films aluminium foils or polythene bags	
	Heat sealer	47.1
	Wrapping equipment	47.3

Beating

Beating is a process which controls the process of crystallization and produces crystals of a small size. For example in the production of fudge, the mass is poured onto the table, left to cool, and then beaten with a wood or metal beater.

Forming/setting

There are two main ways of forming sweets: cutting into pieces, or setting in moulds.

Moulds may be as simple as a greased and lined tray. Other moulds can be made from rubber, plastic, metal, starch, or wood. It is possible to make starch moulds by preparing a tray of cornstarch (cornflour), not packed too tightly. Impressions are then made in the starch using wooden shapes. The mixture is poured into the impressions and allowed to set.

Packaging

When sweets are stored without proper packaging, especially in areas of high humidity, the sucrose may crystallize, making the sweet sticky and grainy. Traditional packaging materials such as banana or sugar-cane leaves are often used to wrap sweets. However, these do not provide sufficient protection for a long shelf-life because they are not efficient barriers to moisture and cannot be securely sealed.

Alternatively, individual wraps can be made from waxed paper, aluminium foil, and cellulose film, or a combination of these. In most cases, the sweets will be wrapped by hand, but for higher production, semi-automatic wrapping machines are available. For further protection, the individually-wrapped sweets may be packed in a heat-sealed polythene bag. Sweets can also be packaged in glass jars, or tins with close fitting lids.

Suitability for Small-scale Production

Certain types of sweets such as hard-boiled sweets require good quality ingredients (such as white granulated sugar). Such ingredients often need to be imported from other parts of the country, as they may not be widely or cheaply available in all areas.

It is possible to produce high-quality sweets on a small scale using inexpensive pieces of equipment. However, an open boiling pan gives less control over the boiling process and ultimately less control over the quality of the finished product. If simple equipment is used, the process will rely heavily upon the producer's experience and skills in production.

By using the basic principles of sweetmaking, the producer can use her/his creative skills to produce a wide range of products from local resources and materials.

Cocoa Butter and Milk Fat

Cocoa butter, also called theobroma oil, is a pale-yellow, edible vegetable fat extracted from the cocoa bean. It is used to make chocolate, as well as some ointments, toiletries, and pharmaceuticals. Cocoa butter has a cocoa flavour and aroma.

Extraction and Composition

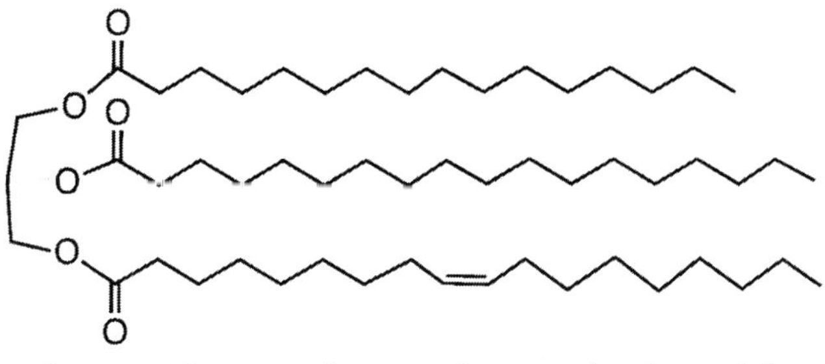

The main constituent of cocoa butter is the triglyceride (fat) derived from palmitic acid, stearic acid, and oleic acid. Cocoa butter is obtained from whole cocoa beans, which are fermented, roasted, and then separated from their hulls. About 54–58% of the residue is cocoa butter. Chocolate liquor is pressed to separate the cocoa butter from the cocoa solids. The Broma process is used to extract cocoa butter from ground cacao beans. Cocoa butter is sometimes deodorized to remove strong or undesirable tastes.

Cocoa butter contains a high proportion of saturated fats, derived from stearic and palmitic acids. Cocoa butter, unlike cocoa solids, has no more than trace amounts of caffeine and theobromine.

Typical Fatty Acid Composition (%):	
Fatty acid	***Percentage***
Arachidic acid (C20:0)	1.0%
Linoleic acid (C18:2)	3.2%
Oleic acid (C18:1)	34.5%
Palmitic acid (C16:0)	26.0%
Palmitoleic acid (C16:1)	0.3%
Stearic acid (C18:0)	34.5%
Other Fatty Acids	0.5%

Adulterants

Some food manufacturers substitute less expensive materials such as vegetable oils and fats (fillers and over-sized packaging) in place of cocoa butter. Several analytical methods exist for testing for diluted cocoa butter. Adulterated cocoa butter is indicated by its lighter colour and its diminished fluorescence upon ultraviolet illumination. Unlike cocoa butter, adulterated fat tends to smear and have a higher non-saponifiable content.

Substitutes

Cocoa butter is becoming increasingly expensive. Substitutes have been designed to use as alternatives. In the United States, 100% cocoa butter must be used for the product to be called chocolate. The EU requires that alternative fats not exceed 5% of the total fat content.

Fats Used for This Purpose Include: Cocoa Butter Substitute – coconut oil or palm oil; Cocoa butter Replacer – soybean oil, rapeseed oil and cottonseed oil; Cocoa Butter Equivalent – shea butter, illipe, sal nut, palm, mango kernel fat, palm oils, and PGPR.

Uses

Cocoa butter is a major ingredient in practically all types of chocolates (white chocolate, milk chocolate, but also dark chocolate). This application continues to dominate consumption of cocoa butter.

Pharmaceutical companies heavily use cocoa butter's physical properties. As a nontoxic solid at room temperature that melts at body temperature, it is considered an ideal base for medicinal suppositories.

Personal Care

Cocoa butter is one of the most stable fats known, a quality that, coupled with natural antioxidants, prevents rancidity, giving it a storage life of two to five years. The velvety texture, pleasant fragrance and emollient properties of cocoa butter have made it a popular ingredient in products for the skin, such as soaps and lotions.

The moisturizing abilities of cocoa butter are frequently recommended for prevention of stretch marks in pregnant women, treatment of chapped or burned skin and lips, and as a daily moisturizer to prevent dry, itchy skin. Cocoa butter's moisturizing properties are also said to be effective for treating mouth sores. However, the largest clinical study regarding the effects of cocoa butter on stretch marks in pregnant women found that results were no different from a placebo.

Physical Properties

The most common form of cocoa butter has a melting point of around 34–38 °C (93–101 °F), rendering solid chocolate at room temperature that readily melts once inside the mouth. Cocoa butter displays polymorphism, having α, γ, β', and β crystals, with melting points of 17, 23, 26, and 35–37 °C respectively. The production of chocolate typically uses only the β crystal for its high melting point. A uniform crystal structure will result in smooth texture, sheen, and snap. Overheating cocoa butter converts the structure to a less stable form that melts below room temperature. Given time, it will naturally return to the most stable β crystal form. Advantage is taken of this phenomenon in the polymorphic transformation theory of chocolate bloom. It is based on the fact that bloomed chocolates are always found to contain the most stable polymorph of cocoa butter. According to this theory, bloom occurs through the uncontrolled polymorphic transformation of cocoa butter from a less stable form to the most stable form.

White Chocolate

White chocolate is a chocolate derivative. It commonly consists of cocoa butter, sugar and milk solids and is characterized by a pale yellow or ivory appearance. The melting point of cocoa butter, its primary cocoa bean component, is high enough to keep white chocolate solid at room temperature.

Composition

White chocolate is not chocolate in the strict sense as it does not contain cocoa solids, the primary nutritional constituent of chocolate liquor. During the manufacturing process, the dark-coloured solids of the cocoa bean are separated from its fatty content (as with milk, semi-sweet, and dark chocolate) but, unlike conventional chocolates, the cocoa solids are not later recombined. As a result, white chocolate contains only trace amounts of the stimulants theobromine and caffeine, while lacking the antioxidant properties or many characterizing ingredients of chocolate, such as thiamine, riboflavin, and phenylethylamine. Often, the cocoa butter is deodorized to remove its strong flavour.

Some preparations known as confectioner's coating or summer coating may be confused with white chocolate, but are made from inexpensive solid or hydrogenated vegetable and animal fats, and are not at all derived from cocoa. These preparations may actually be white (in contrast to white chocolate's ivory shade) and will lack cocoa butter's flavour.

Regulations

Regulations govern what may be marketed as "white chocolate": In the United States, since 2004, white chocolate must be (by weight) at least 20% cocoa butter, 14% total milk solids, and 3.5% milk fat, and no more than 55% sugar or other sweeteners. Before this date, U.S. firms required temporary marketing permits to sell white chocolate. The European Union has adopted the same standards, except that there is no limit on sugar or sweeteners.

Symmetrical SOS-type CBAs: Cocoa Butter Equivalents (CBEs)

Cocoa Butter (CB) is an important constituent of chocolate formulation. It enjoys a very high status in the chocolate industry due to its property of remaining hard, moldable and brittle at room temperature and completely melting at the mouth temperature.

However, CB has its own disadvantages such as:

- Low milk-fat tolerance.
- Lack of stability at elevated temperatures.
- Tendency to bloom.

Efforts were then made to find certain fats produced from tree borne oilseeds that have unique triglyceride combination consisting of the right proportion of POP, POSt and StOSt types of glycerides. These fats matched the properties of cocoa butter and could be blended with cocoa butter proportionately.

So, basically Cocoa Butter Equivalents (CBEs) are fats, which behave like cocoa in all respects and are able to mix with cocoa butter in any proportion without altering the melting rheological and processing characteristics of cocoa butter in all types of formulation. CBEs are non-mono-unsaturated triglycerides like cocoa butter.

Palm Oil is fractionated to produce middle-melting fractions rich in POP. So also, exotic fats such as Sal, Mango, Shea, Kokum etc. are fractioned to get triglyceride cuts in POS and SOS . A careful preparation and blending of these results in tailor-made fat that is equivalent to cocoa butter.

The Principle Advantages of Incorporating CBEs are:

- Reduction in the production cost of chocolate, as CBE's are cheaper than cocoa butter.
- Stabilizing influence on fluctuating price of cocoa butter.
- Improves tolerance of milk-fat.
- improves stability of chocolates in tropical climate.
- Last nut not the least ,bloom control.

Demand of Cocoa Butter Equivalents:

Other than chocolates, these fats are also predominantly used for the production of compound coatings used for enrobing bakery and confectionary products where the coating forms 10% to 30% of the total weight of the enrobed products.

Adulteration and its Detection

Adulteration, the act of debasing a pure or genuine commodity for pecuniary profit, by adding to it and inferior or spurious article, or by taking from it one or more of its constituents. The term is derived from the Latin adultero, which in its various inflections signifies to defile, to debase, to corrupt, to sophisticate, to falsify, to counterfeit. The objects of adulteration are fourfold, namely, to increase the bulk or weight of the article, to improve its appearance, to give it a false strength, or to rob it of its most valuable constituents.

All these adulterations are manifestly of a designedly fraudulent character, and are therefore probably the subjects of judicial enquiry; but there may be accidental corruptions and adulterations of a commodity, arising from natural or unavoidable causes, as when darnel or ergot become mixed with grain in the fields of the slovenly farmer, or when an article becomes changed and deteriorated from spontaneous decay, or when mineral matters and other impurities are accidentally derived from the machinery or vessels in which the thing is prepared or kept.

The recognition of such impurities, and the tracing of them to their source, is of prime importance in pursuing a charge of adulteration. Few articles of commerce, however, are exempt from fraudulent adulteration, and the practice of it has grown with the competition of trade, and the removal of those wholesome restrictions which in former times were so energetically opposed to all kinds of dishonest dealing; for the guilds and companies of all large cities had their corporate regulations for supervising and governing every description of trade and manufacture.

The excise, too, including the customs, had until recently control over the quality of all exciseable articles; and although the prime object of this was to protect the revenue of the country, yet it also served to prevent adulteration. In addition to this there were in ancient times ordinances of assize for regulating the price and quality of the common necessaries of life.

Many common foods contain foreign materials introduced with the direct object of defrauding the public and securing a larger profit to the seller. Even the ordinary food-stuffs of the breakfast-table are not always what they seem, with the exception, perhaps, of sugar, on the purity of which one can depend.

In the case of milk the chief, and one might almost say the natural, adulterant is water. New milk contains as much as 87 per cent, of water, and the uninitiated might suppose that it would be very easy to add a little more without detection. Careful analysis, however, will always reveal any such manipulation, although it must be borne in mind that there may be a certain difference in the richness of milk from various cows.

One method which the chemist has at his disposal is the determination of the specific gravity - that is, he finds out how much heavier the milk is than an equal bulk of water.

If a bulk of water were taken which weighed exactly 100 ounces, an equal bulk of pure new milk would weigh about 103 ounces, a little less or a little more, according to its source. That is, the average specific gravity of milk may be taken as 1.03. If, then, a certain sample of milk had a specific gravity of only 1.02, we might be sure that it had been "watered." On the other hand, the fact that the specific gravity of a sample is 1.03, does not prove the milk to be satisfactory; for, curiously enough, it is possible, by a judicious combination of watering and skimming, to get a product which has the same specific gravity as the original milk.

Butter is another household article that is readily and frequently adulterated, although the recent Butter and Margarine Act should do something to protect the public. The usual frauds practised in the case of butter are (1) the sale of "renovated" or "process" butter as fresh butter, and (2) the substitution of a certain amount of cheap beef fat or lard for the true butter fat. Renovated butter is obtained from rancid butter by a process in which the objectionable matter is removed; the product is rendered sweet for the time being, and is sold as choice butter.

Artificial butter, on the other hand, or margarine, as it is commonly called, is prepared from beef fat or lard, which is worked up with ordinary butter and colouring matter so as to resemble the real article. Besides a certain difference in the taste of butter and margarine, there is one very simple method, known as the spoon test, by which they may be distinguished. If a little genuine butter is melted in a large spoon over a small Bunsen flame, and the heating is continued, the butter ultimately boils quietly and foams up to the edge of the spoon. Margarine, treated in the same way, splutters about and crackles, but does not foam. The practice of selling margarine under the name of pure butter is probably dying out, but it is not so very long ago since a bold individual was prosecuted for actually advertising a process for the "scientific" blending of butter with beef fat or lard. Science, it would seem, covers a multitude of sins.

A food-stuff which is very frequently adulterated is chocolate. This substance is obtained by grinding cocoa nibs, which are the crushed kernels of cocoa beans. The nibs consist to about 45 per cent, of a fat, the so-called cocoa butter, and in this respect are quite different from the shells of the cocoa bean, which contain only 2 to 3 per cent, of the fat. Seeing that the price of cocoa nibs is about ten times that of cocoa shells, the common practice of adulterating chocolate with powdered cocoa shells is distinctly profitable. This fraud is best detected by the aid of the microscope, an instrument which is part of the necessary equipment of an analytical chemist's laboratory. To the practised eye the presence of the powdered shells is at once obvious.

There is another adulterant of chocolate or cocoa which is easily detected with the aid of the microscope, and that is starch. This substance is very widely distributed in the plant world, and occurs in all sorts of vegetables and cereals. The samples of starch obtained from these various sources, such as wheat, rice, potatoes, and maize, are chemically identical, but when they are examined under the microscope, the granules of which they consist are found to be surprisingly different in size and shape. The granules of wheat starch are circular, those of potato starch are oval, while those of rice starch are many-sided; the granules from maize starch, as found, for example, in corn-flour, are also many-sided, but are uniformly much larger than rice-starch granules. It is therefore possible for a skilled analyst to determine with the microscope whether any starch, and, if so, what kind of starch, has been used in adulterating a given food-stuff. He can also discover at once whether a certain kind of starch is pure or is contaminated with another kind. Obviously there is a temptation for the adulterator to add a cheap starch to a more expensive one, say potato starch to arrowroot, keeping the price the same. The microscope, however, soon exposes such a fraud.

Chocolate & Chocolate Liquor - Adulteration with Insect and Rodent Filth

Regulatory Action Guidance: The following represents criteria for direct reference seizure to the Division of Compliance Management and Operations (HFC-210), and for direct citation by District Offices:

1. Insect Filth

a. The chocolate in six (6) 100 gram subsamples contains an average of 60 or more insect fragments per 100 grams.

or

b. Any one subsample contains 90 or more insect fragments, even if the overall average of all the subsamples is less than 60.

2. Rodent Filth

a. The chocolate in six (6) 100 gram subsamples contains an average of more than 1.0 rodent hair per 100 grams, regardless of the size of the hairs or hair fragments.

or

b. Any one subsample contains more than 3 rodent hairs even if the overall average is less than 1.0 rodent hair.

NOTE: Insect and rodent filth analyses are conducted using method 965.38, Official Methods of Analysis, (2000) 17th Edition, AOAC, Gaithersburg, MD.

Moulded bar and Coating Applications

Chocolate molds have been around since chocolate consumption moved from predominately drinking chocolate to predominately eating chocolate. In order to produce bite-size chocolate pieces for eating, the large blocks of chocolate have to be molded into smaller forms.

In the late 1800s, creating fanciful chocolate molds of metal was at its height. Small bakers and chocolatiers produced intricate shapes, both flat and three-dimensional. The introduction of plastic molds revolutionized both the industrial and home markets. Now plastic molds, usually based on historical designs first done in metal, can be produced inexpensively.

When purchasing chocolate molds, look for strong plastic with deep intricate designs. These will produce a much more elegant product than the shallow, less-detailed molds you'll usually find in craft or kitchen stores. The molds should never be washed in soap. Just use hot water. Soap scum can mar the taste of the chocolate if not thoroughly removed. If your chocolate is tempered properly, most of the chocolate can be flaked off with a soft cloth.

After the finished chocolate is removed from the mold, it leaves a light film of cocoa butter which makes the next chocolate made in the mold gleam even more.

The Molding Process

You'll need to have mastered tempering chocolate before you attempt to mold with it. Put out your molds, utensils (ladle, spoons, pastry scraper or pallet knife, piping bag) and parchment paper or silpat mat before you temper your chocolate so you can start molding right away.

Try to buy at least two of each kind of mold as this allows you to work on one mold while the other is hardening. When molding, you want to work as quickly as possible and remember to stir your tempered chocolate a lot during the process as it will tend to loose its temper and start to solidify if left alone. Remember that chocolate hates to be ignored.

Step 1: Fill the Mold

Quickly fill your mold cavities using either a ladle or spoon. Don't worry if you have chocolate all over the mold as you're going to scrape off all the excess later. Once the cavities are full, tap the mold briskly on the table to settle the chocolate into any grooves or designs in the cavity and to force out air bubbles.

Step 2: For Solid, Molded Chocolate Shapes

Scrape off the excess using a pastry scraper or pallet knife. Tap the mold on the table again to settle the chocolate into the cavities. Place the mold (cavity side up) on a flat cool surface to harden.

A note about achieving perfect detail if your mold cavity is very detailed: To reduce the risk of small bubbles marring the mold detail, you can paint the mold cavities with a thin coat of tempered chocolate using a soft brush before proceeding to fill the molds.

Step 2: For Shell Molds to Be Filled with Your Truffle or Candy Mixture

Quickly fill your mold cavities using either a ladle or spoon and then turn the mold over your bowl of chocolate and let the excess chocolate run out of the cavities, leaving a thin coating of chocolate. Tap the bottom of the mold gently to release more chocolate during this process.

Turn the mold upright and scrape off the excess across the surface of the mold using a pastry scraper or pallet knife. A little of the scraped chocolate might run into the cavities but that is fine. Turn the mold over again and place it (cavity side up) on parchment paper or silpat mat to harden.

Filling the chocolate cavity: Make your favourite chocolate truffle recipe and have it in a plastic or parchment piping bag. When the chocolate-lined cavities have hardened, pipe the mixture into the cavity making sure not to fill more than 3/4 to 7/8 of the way up the sides of the cavity. Try hard not to get any filling on the outer rim of the chocolate. This is where the piping bag works great.

If you do get some filling on the edge of the chocolate, scrape it off immediately because it will keep your chocolate bottom from adhering properly. When all the cavities are filled, set the filled shells aside for a few minutes to let the filling settle and form a slightly dry skin on the top.

When the shells are ready, ladle chocolate across the top of the mold as before and then use a pastry scraper or pallet knife to quickly scrape off the excess chocolate. Place the filled mold (cavity side up) on a flat, cool surface to harden.

Step 3: Hardening Your Molds

If you don't have a cool room or cupboard in which to harden the chocolate in the molds or are pressed for time, you can use the refrigerator to speed up the process. Set aside one shelf in the fridge and have it empty before you start tempering the chocolate and molding. Once the mold is full, put it on the shelf and set a timer for five minutes. Remove the mold from the refrigerator and let it continue hardening at cool room temperature.

The chocolates are ready to be released when the chocolate pulls away from the mold slightly and the outside of the cavity appears grayish if you look at the bottom of the mold. This is an indication that air is being pulled in between the chocolate and the mold as the chocolate contracts and hardens.

Once you think the mold is ready, turn it over on parchment paper or a Silpat mat. If your chocolate is tempered properly and it is completely hardened, the chocolates should just fall out of the mold. If they don't, give the mold a little tap on the bottom to help release the chocolates. If that doesn't work, then set them aside and check later.

You can return them to the refrigerator, but again, not for more than three minutes — so set the timer. The problem with leaving the filled molds in the refrigerator for too long is that the cold environment will start to pull the sugar in the chocolate itself to the surface, which is up against the mold. The sugar will adhere to the mold and make it very difficult to remove the chocolates.

Chocolate Bar

A chocolate bar is a confection in bar form comprising some or all of the following components: cocoa solids, cocoa butter, sugar, and milk. The relative presence or absence of these components form the subclasses of dark chocolate, milk chocolate, and white chocolate. In addition to these main ingredients, it may contain emulsifiers such as soy lecithin and flavours such as vanilla.

In most of the English-speaking world, chocolate bar also refers to what is typically called a candy bar in American English. This is a form of confectionery usually packaged in a bar or log form, often coated with chocolate and sized as a snack for one person. But within that term, a wide variety of products exist, ranging from solid chocolate bars to multiple layerings or mixtures of ingredients such as nuts, fruit, caramel, or fondant containing no chocolate.

Certain brands of chocolate bars and candy bars are sold as being for nutritional supplementation purposes. These bars contain protein and various vitamins while still retaining a sweet taste.

Compound Chocolate

Compound chocolate is a product made from a combination of cocoa, vegetable fat, and sweeteners. It is used as a lower-cost alternative to true chocolate, as it uses less-expensive hard vegetable fats such as coconut oil or palm kernel oil in place of the more expensive cocoa butter. It may also be known as 'compound coating' or 'chocolatey coating' when used as a coating for candy.

Often used in lower-grade candy bars, compound chocolate is designed to simulate enrobed chocolate on a product.

Cocoa butter must be tempered to maintain gloss and coating. A baker tempers chocolate by cooling the chocolate mass below its setting point, then rewarming the chocolate to between 31 and 32 °C (88 and 90 °F) for milk chocolate, or between 32 and 33 °C (90 and 91 °F) for semi-sweet chocolate. Compound coatings, however, do not need to be tempered. Instead, they are simply warmed to between 3 and 5 °C (5.4 and 9.0 °F) above the coating's melting point.

Filling Applications

Fillings are a valuable component of many sweets. Vegetable fats represent an important recipe constituent. Fat forms the continuous phase, in which the other components are embedded. That's why physical and chemical properties of fat are decisive for the quality of the finished product. Factors like texture, mouth feel and flavour release are determined by the fat.

For these fats melting properties are particularly important: the fats have to melt rapidly and completely in the mouth without leaving an unpleasant fatty impression on the tongue.

The desired consistency of the filling limits the choice of suitable fats. Especially fillings for chocolate and pralines contain high amounts of other fats like milk fat, hazelnut and almond oils. Therefore one has to bear in mind that the consistency of the filling softens naturally. Corresponding harder filling fats have to be used in order to achieve the desired consistency of the filling.

Using Specialty Fats for Fillings on Trend

Manufacturers looking to expand their share of the filled chocolate market are keen to explore new ingredients and efficient processes capable of giving consumers the exciting, moreish sensation they expect every time they eat a piece of snack. A good place to start that search is with the fat – around 30% of every filling is the filling fat, which has a key influence on the stability, melting properties, texture and flavour release.

Vegetable fats often hold the key to meeting consumer requests for new, indulgent products. One strong indulgence trend favours the special cooling sensation that gives filled chocolates a fresher feel and accentuates flavours – a sensation closely linked to the vegetable filling fat and influenced by the process, overall fat composition and the recipe as a whole.

At the same time, demand is growing for healthier products with a lower content of saturated fat per serving. This fat is an important structure builder, so developing a more indulgent, specialty filling with less saturated fat is a challenge. However, nothing is impossible provided that the optimal vegetable fat is identified.

The Fats in the Filling

Generally, 35% to 50% of a fat-based chocolate filling is a mixture of vegetable fat, milk fat, cocoa butter, nut oil and oils from other ingredients. While vegetable fat dominates the characteristics of the total fat phase, it is still influenced by the other fats that are present. This is why the choice of vegetable fat is so important to achieving the desired sensory quality, shelf life and process behaviour.

For indulgent fillings, specialty fats can contribute stable aeration during processing, add an appealing melting profile and give impeccable bloom stability. They can also give a differentiating cooling effect. Specialty fats for healthier fillings must be free of trans fatty acids and non-hydrogenated. Equally important, they must enable a reduction in saturated fats without compromising the texture.

A Cool Sensation

Cool-melting filling fats belong to one of two triacylglycerol groups (TAGs):

I. A lauric-based oil made from palm kernel oil and/or coconut oil or their fractions, or

II. An oil blend dominated by symmetric monounsaturated triacylglycerols

Both groups are characterized by a high content of the same or similar melting TAGs. These interact with the rest of the fats in the recipe to form a narrow melting profile, which melts primarily below 30°C.

Table shows the close connection between the fat phase and cooling sensation. The fat phase in the truffle recipe seems to have the right composition to obtain the fast melting properties that secure a high cooling sensation – a cooling index of 9.6.

Here, vegetable fat A has been mixed with a high content of cocoa butter and diluted with nut oil. In the yoghurt recipe, however, vegetable fat A reduces the cooling sensation by around 35 %. When vegetable fat B is used instead of vegetable fat A, the cooling sensation in the same yoghurt recipe is recovered.

Table 1. All three fillings were produced in the same way and evaluated by a sensory panel after one week of storage at 20°C.

	Fat composition		
Recipe	Truffle	Yoghurt	Yoghurt
Veg. fat A	52.5	95.0	0.0
Veg. fat B	0.0	0.0	95.0
Milk fat	4.0	5.0	5.0
Hazelnut oil	19.5	0.0	0.0
Cocoa butter	24.0	0.0	0.0
Total fat content	52.0	52.0	52.0
Cooling index	9.6	6.0	9.9

This clearly illustrates the importance of the vegetable fat in optimizing the cooling sensations. Total fat impact The cooling sensation occurs when the fat melts and heats up from room temperature to body temperature. In the mouth, specialized sensory receptors – thermoreceptors – code absolute and relative changes in temperature. The faster the melting process and the more energy required, the more distinctive the cooling sensation experienced by the consumer. The total fat content at room temperature is, therefore, directly connected to the potential for a cooling sensation in the mouth, where the filling melts almost entirely.

The speed and ease with which a filling can be melted and manipulated in the mouth depends on the properties of the suspension. Low viscosity and, especially, a low yield value, created by the addition of polyglycerol polyricinoleate (PGPR), speed up the distribution of the filling in the mouth, increasing the surface area. This, in turn, increases the melting rate and the number of activated thermoreceptors.

Filling processing The final cooling sensation also depends on how the filling is cooled during processing, whether based on a group I or group II vegetable fat. The optimal cooling temperature of a group I fat is 5-10°C, while a group II fat is best cooled at 10-15°C. Because of the impact of the final fat composition, application and air velocity in the cooling tunnel, the optimal temperature must be evaluated for each application and filling.

As group II fats are rich in polymorphic TAGs, they must be tempered or pre–crystallized before use. If not, they will crystallize in a lower polymorphic form and, consequently, transform to a higher form during storage, with a significant negative impact on the cooling sensation.

Lowering Saturates

Altering the saturated fat without changing texture or melting properties is a very big challenge in the development of healthier confectionery products. It is important to look beyond the saturated fatty acid content of the vegetable fat to the saturated fats in the entire fat phase. Maintaining all the characteristics of a filling is very difficult if the fat composition undergoes considerable change.

Opportunities for Altering the Total fat Composition Include

1. Adjusting the ratio between different components, for example by incorporating more wafer, nuts and filling and using less coating
2. Replacing fats from ingredients and substituting with a vegetable fat, giving more structure for each % of saturated fatty acid
3. Slightly lowering the total fat content
4. Replacing the vegetable fat with another that contains less saturated fat but has the same texture.

In most cases, the best approach is to start by optimizing the first three points. This will ensure that the final solution is closer to the original product in terms of taste and production settings.

It Will also Maximize the Reduction in Saturated Fat.

Fats with different textures Different fats produce different textures, even though their content of saturated fat is the same. This is important to bear in mind when reducing saturated fats. A main influencing factor is the chain length of the fatty acids – a shorter chain length has a lower melting point.

One example is milk fat which, despite its content of saturated fat, contributes very little structure.

Figure below shows how texture can vary from fat to fat. All measurements are based on the same filling recipe with the same particle size and the optimal process for each fat system. The only variant is the type of fat used.

Figure: *Texture analysis using a Texture Analyser TX21 with needle P/2N after one week of storage at 20° C. Penetration:5mm.*

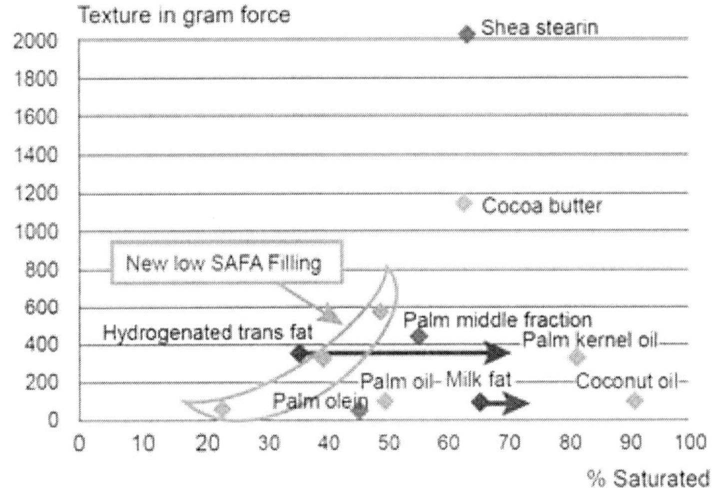

The two red points indicate the presence of trans fatty acids. These produce a texture similar to saturated fats and, consequently, much higher than indicated by their saturated fat content.

The three green points are all examples of low saturated filling fats, where the target is to maximize the texture obtained from the saturated fat. These vary from very soft to semi-hard, covering most of the existing filling range today. They are all non-hydrogenated and do not require tempering, making them easy to use.

The four yellow points are examples of well-known confectionery fats, while the three blue points show three fractions of well-known fats. 25% reduction in saturates These texture analyses correlate perfectly with the findings of a sensory panel regarding filling hardness, whereas the correlation with solid fat content varies from fat to fat.

They also show that it is possible to cut saturated fats by 25% by switching from milk fat to palm oil, for example, without any significant change in texture at 20°C. An even higher reduction can be achieved by replacing milk fat with one of the new low saturated fats. Fat interactions A typical filling comprises several fats which interact in different ways. One well-known example is the combination of lauric fat and cocoa butter. Although the content of saturated fat is unchanged when the two fats are mixed, the incompatibility of the two fats makes the mixture much softer.

Such fat interactions are important to consider when developing new low saturated fillings. The compatibility of vegetable fats with milk fat and cocoa butter is particularly important because of the wide use of these fats in filling recipes.

New Opportunities

Optimizing a filling is easier said than done. But when it is done right, there are many opportunities to bring new, exciting products to the market – in line with today's trends. Choosing the right vegetable filling fat is the first step towards a fresh, cool-melting filling or a healthier filling reduced in saturated fat. With the appearance of new vegetable fat solutions, it has become possible to produce new indulgent fillings that overcome texture challenges with comparative ease.

Caramel

Caramel is a beige to dark-brown confectionery product made by heating any of a variety of sugars. It is used:

- as a flavouring in puddings and desserts
- as a filling in bonbons
- as a topping for ice cream and custard

The process of caramelization consists of heating sugar slowly to around 170 °C (340 °F). As the sugar heats, the molecules break down and re-form into compounds with a characteristic colour and flavour.

A variety of candies, desserts, and confections are made with caramel: brittles, nougats, pralines, crème brûlée, crème caramel, and caramel apples. Ice creams sometimes are flavoured with or contain swirls of caramel.

Caramel Sauce

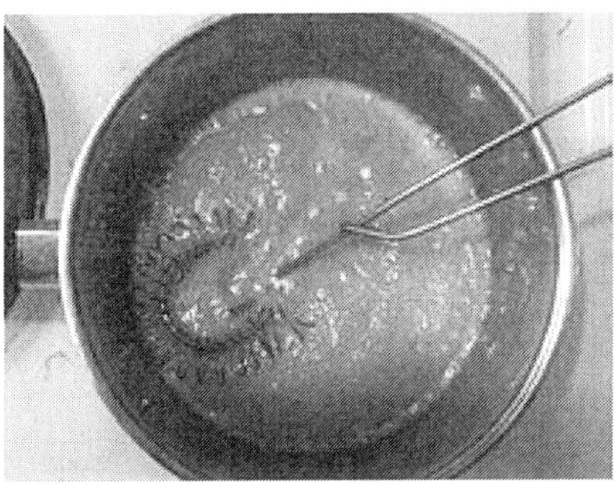

Figure: *Caramel sauce being made in a pan with a whisk.*

Caramel sauce is made by heating water and caster sugar (also called superfine sugar) at a low to moderate temperature until the sugar dissolves and "caramelizes," changing colour to golden brown.

Figure: *Milk caramel sold as square candies, either for eating or for melting down.*

Caramel Candy

Caramel candy is a soft, dense, chewy candy made by boiling a mixture of milk or cream, sugar(s), butter, and vanilla flavouring. The sugar(s) are heated separately to reach 170 °C (340 °F), caramelizing them before the other ingredients are added. Alternatively, all ingredients may be cooked together; in this procedure, the mixture is not heated above the firm ball stage (120 °C (250 °F)), so that caramelization of the milk occurs but not caramelization of the sugars. This type of candy is often called milk caramel or cream caramel.

Caramel Colouring

Caramel colouring, a dark, bitter-tasting liquid, is the highly concentrated product of near total caramelization, bottled for commercial use. It is used as food colouring and in beverages such as cola.

Chemistry

Caramelization is the removal of water from a sugar, proceeding to isomerization and polymerization of the sugars into various high-molecular-weight compounds. Compounds such as difructose anhydride may be created from the monosaccharides after water loss. Fragmentation reactions result in low-molecular-weight compounds that may be volatile and may contribute to flavour. Polymerization reactions lead to larger-molecular-weight compounds that contribute to the dark-brown colour.

In modern recipes and in commercial production, glucose (from corn syrup or wheat) or invert sugar is added to prevent crystallization, making up 10%–50% of the sugars by mass. "Wet caramels" made by heating sucrose and water instead of sucrose alone produce their own invert sugar due to thermal reaction, but not necessarily enough to prevent crystallization in traditional recipes.

Nutritional Information

Two tablespoons (i.e., 41 grams) of commercially prepared butterscotch or caramel topping contain:

- Calories (kcal): 103
- Protein (g): 0.62
- Total lipids (fat): 0.04
- Carbohydrates, by difference (g): 27.02
- Fibre, total dietary (g): 0.4
- Cholesterol (mg): 0.0

Truffles

A truffle is the fruiting body of a subterranean Ascomycete fungus, predominantly one of the many species of the genus *Tuber*. Some of the truffle species are highly prized as a food. French gourmand Jean Anthelme Brillat-Savarin called truffles "the diamond of the kitchen". Edible truffles are held in high esteem in French, Georgian, Greek, Italian, Middle Eastern and Spanish cooking, as well as in international *haute cuisine*.

Truffles are ectomycorrhizal fungi and are therefore usually found in close association with the roots of trees. Spore dispersal is accomplished through fungivores, animals that eat fungi.

The mycelia of truffles form symbiotic, mycorrhizal relationships with the roots of several tree species including beech, poplar, oak, birch, hornbeam, hazel, and pine. They prefer argillaceous or calcareous soils that are well drained and neutral or alkaline. Truffles fruit throughout the year, depending on the species and can be found buried between the leaf litter and the soil.

Phylogeny

The phylogeny and biogeography of truffles was investigated in 2008 using internal transcribed spacers (ITS) of nuclear DNA with five major clades (Aestivum, Excavatum, Rufum, Melanosporum and Puberulum); this was later improved and expanded in 2010 using large-subunits (LSU) of mitochondrial DNA to nine major clades. The Magnatum and Macrosporum clades were distinguished as distinct from the Aestivum clade. The Gibbosum clade was resolved as distinct from all other clades, and the Spinoreticulatum clade was separated from the Rufum clade.

Types

White Truffle

Figure: *White truffle washed and cut*

The "white truffle" or "trifola d'Alba Madonna" (*Tuber magnatum*) comes from the Langhe and Montferrat areas of the Piedmont region in northern Italy and, most famously, in the countryside around the cities of Alba and Asti; in Italy it can also be found in Molise, Abruzzo, and in the hills around San Miniato, in Tuscany. It is also found on the Istria peninsula, in Croatia in the Motovun forest along the Mirna river, and in Slovenia along the Dragonja and Rizana river, as well as in the Drome area in France.

Growing symbiotically with oak, hazel, poplar and beech and fruiting in autumn, they can reach 12 cm (5 in) diametre and 500 g, though are usually much smaller. The flesh is pale cream or brown with white marbling. Italian white truffles are very highly esteemed (*illustration, left*) and are the most valuable on the market: The white truffle market in Alba is busiest in the months of October and November when the Fiera del Tartufo (truffle fair) takes place. In 2001, the *Tuber magnatum* truffles sold for between \$1000–\$2200 per pound (\$2000–\$4500 per kg); as of December 2009 they were being sold at \$14,203.50 per kilogram.

In 1999, Giancarlo Zigante and his dog Diana found one of the largest truffles in the world near Buje, Croatia. The truffle weighed 1.31 kilograms (2 lb 14 oz) and has entered the *Guinness Book of Records*.

The record price paid for a single white truffle was set in December 2007, when Macau casino owner Stanley Ho paid \$330,000 (£165,000) for a specimen weighing 1.5 kilograms (3.3 lb), discovered by Luciano Savini and his dog Rocco. One of the largest truffles found in decades, it was unearthed near Pisa, Italy and sold at an auction held simultaneously in Macau, Hong Kong and Florence. This record was then matched on November 27, 2010 when Ho again paid $330,000 for a pair of white truffles, including one weighing nearly a kilogram.

In December 2014, a White Alba's Truffle weighing in at 4.16 pounds or 1.89 kilos was unearthed in the Umbrian region of Italy by the Balestra Family of Sabatino. It was auctioned at Sotheby's in New York. While some had expected it to sell for $1 million, it was sold for $61,000 to a Taiwanese buyer.

The *Tuber magnatum pico* white truffle is found mostly in northern and central Italy, while the *Tuber borchii*, or whitish truffle, is found in Tuscany, Abruzzo, Romagna, Umbria, the Marche and Molise. Neither of these is as aromatic as those from Piedmont, although those from Città di Castello come quite close.

Black Truffle

The black truffle or black Périgord truffle (*Tuber melanosporum*), the second-most commercially valuable species, is named after the Périgord region in France and grows with oak and hazelnut trees. Black truffles are harvested in late autumn and winter. The genome sequence of the black truffle was published in March 2010.

Summer or Burgundy Truffle

The black summer truffle (*Tuber aestivum*) is found across Europe and is prized for its culinary value. Burgundy truffles (*Tuber uncinatum*) are harvested in autumn until December and have aromatic flesh of a darker colour.

Other Species

A less common truffle is "garlic truffle" (*Tuber macrosporum*). In the U.S. Pacific Northwest, several species of truffle are harvested both recreationally and commercially, most notably, the "Oregon white truffles", *Tuber oregonense* and *Tuber gibbosum*.

The "pecan truffle" (*Tuber lyonii*) syn. *texense* is found in the Southern United States, usually associated with pecan trees. Chefs who have experimented with them agree "they are very good and have potential as a food commodity".

Although pecan farmers used to find them along with pecans and discard them, considering them a nuisance, they sell for about $100 a pound and have been used in some gourmet restaurants.

Culinary Use

Figure: *Truffle oil (olive oil with* Tuber melanosporum*).*

Because of their high price and their pungent aroma, truffles are used sparingly. Supplies can be found commercially as unadulterated fresh produce or preserved, typically in a light brine.

White truffles are generally served raw, and shaved over steaming buttered pasta or salads or fried eggs. White or black paper-thin truffle slices may be inserted into meats, under the skins of roasted fowl, in *foie gras* preparations, in *pβtés*, or in stuffings. Some speciality cheeses contain truffles, as well.

The flavour of black truffles is far less pungent and more refined than that of white truffles. Its strong flavour is often described as syrupy sweet. Black truffles also are used for producing truffle salt and truffle honey.

While in the past chefs used to peel truffles, in modern times, most restaurants brush the truffle carefully and shave it or dice it with the skin on so as to make the most of this valuable ingredient. A few restaurants, such as Philippe Rochat in Switzerland, still stamp out circular discs of truffle flesh and use the skins for sauces.

Truffle Oil

Truffle oil is used as a lower-cost and convenient substitute for truffles, to provide flavouring, or to enhance the flavour and aroma of truffles in cooking. Most "truffle oil", however, does not contain any truffles. The vast majority is olive oil which has been artificially flavoured using a synthetic agent such as 2,4-dithiapentane.

Food Coating and Fat Migration

Coating is an industrial process that consists of applying a liquid or a powder onto the surface of a product of any possible shape to convey new (usually sensory) properties. Coating designates an operation as much as the result of it: the application of a layer and the layer itself. Coating takes different meanings depending on the industry concerned.

There are many similarities between coating processes and numerous examples of technology transfer to and from the food industry.

Coating in the food industry is the application of a layer of liquids or solids onto a product. The operation essentially relies on mechanical energy. It consists mostly in setting the product particles in motion and simultaneously applying the coating ingredient in a certain pattern to expose one to the other. It involves such phenomena as adhesion, friction, viscosity, surface tension and crystallisation. Food coating is not a "hard" science such as drying or cooling, which can be described by equations and are predictable. Food coating is rather a "soft" knowledge derived from the accumulation of know-how. One reason is that the product and the ingredients considered have complex characteristics, variations and interactions.

Encapsulation is the application of a liquid layer to very small particles. It relies on an array of principles: entrapping a molecule inside a matrix, chemical bonding, and polymerisation. Encapsulation aims at the protection and controlled release of active molecules when immersed in an environment. As a rule of thumb, particle size can discriminate between "encapsulation" (below 300 μm to 1000 μm) and "food coating" (above this limit). Mere mechanical movement is not adequate and sufficient to fulfill the proper coating of minute particles.

Examples of coated products					
Finished product	***Base***	***Ingredient***	***Post treatment***	***Rate, %***	***Objective***
Ready-to-eat cereals	Expanded cereals	Sugar syrup	Drying	20-50%	Flavour and appearance enhancement
Dragees	Nuts, chocolate, sweets	Sugar	Panning	10-100%	Taste, flavour
Chocolate	Hazelnuts, almonds	Chocolate	Cooling	30-50%	Taste, flavour
Prepared vegetable	Frozen vegetables	Water, fats, flavourings	Freezing	15-100%	Taste, convenience
Processed cheese	Grated/shredded cheese	Anticaking	None	<2%	Prevent agglomeration
Nuggets	Meat	Batter and crumb	Frying	30-50%	Palatability, cost, appearance
Snacks	Expanded flour	Oil and seasoning	None	5-40%	Palatability
Crunchy nuts	Peanut	Flour, binder, seasoning	Frying, baking	30-100%	Palatability

Objectives of Coating

- Organoleptic: Coatings can be added for the enhancement of organoleptic properties of a food product. Appearance and palatability can be improved by adding colour (white dragee, brown chocolate), changing the surface aspect (glazed sweets or rough, crispy nuggets); changing or adding tastes (sweet dragee, salted snack) or flavours (fruit-glazed sweet goods), or texture (breaded crispy nuggets).
- Nutrition: Coatings also can be used to add vitamins and minerals (enriched white rice) or food energy.
- Functional: Coating conveys functional properties, such as particle separation (oiled dry fruit, shredded cheese), antioxidant effect (fruit cubes), or a barrier effect [water migration between a layer of ice cream and a biscuit (cookie) or against moisture lost of chewing gum]. Barrier effects are often difficult to achieve.
- Cost: An ingredient may be cheaper than the product it coats and thus allows for a slight cost reduction.

The Coating Process

The coating process begins with the application of the coating on the food product, but the end product must be stable throughout its shelf life. Therefore, a coating process is completed by a stabilizing process, either by freezing, cooling, heating or drying.

Sequences of the Coating Process

Application: To apply minute quantities of an ingredient, spraying is used to disperse it first, instead of just pouring it. This hastens the dispersion on the whole surface of the product.

For larger ratios of coating to substrate, mixing or dipping can be used. Multiple stages also can be used; breaded meats, for example, may have a dry application (predust) followed by a wet batter dip and then another dry crumb application.

- Adhesion: the coating must adhere to the product, meaning there must be a degree of affinity between the ingredient and the product.
- Coalescence: in case of a liquid, the multiple droplets may merge to form a uniform continuous layer. Characteristics of the ingredient in relation to the product, such as viscosity and surface tension associated to a mechanical effect (friction) are critical.
- Stabilisation : depending on the nature of the coating ingredient(s) and substrate product, the ingredient is stabilised by elimination of the solvent (drying and evaporation of water, alcohol), crystallisation (sugar crystallises when water is evaporated, fat crystallises when cooled), or thermal treatment (proteins set irreversibly when heated).

Coating as a System

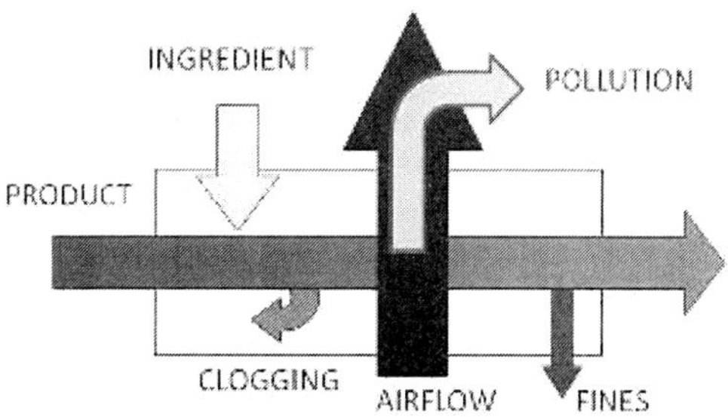

Figure: *A coating process can be broken into the following elements:*

- Inputs: base product, additives and ingredients
- Additional flows: air as a carrier of product or ingredient, or for drying, energy in a mechanical (agitation, transfer, friction) or a thermal form (convection, conduction or radiation heating)
- Outputs: end product, excess of coating ingredient, lost or to be recycled

Collaterals Occur Along the Process

- Breakage of product
- Generation of *fines* (small particles)
- Agglomeration of products
- Clogging of system surfaces with product or ingredient
- Airborne pollution, volatile organic component

These effects generally are to be avoided unless the end product is made more desirable.

The parametres of the system Parametres affecting the system are listed by origin:

Initial characteristics.		
Base product	*End product*	*Production*
Shape, size, distribution, bulk density, nature, surface aspect, resistance, composition, flow behaviour, fines, hygroscopicity, temperature	Capacity, end aspect, weight gain, storage behaviour, resistance	Recipe changes, duration, cleaning

This first set of criteria governs the choice of the coating ingredient. The coating consists either in a single ingredient or a mix. This mix has different physical forms: solution, emulsion, suspension, powder, etc. It has its own characteristics. In addition, a fluid may be required such as spraying, cooling, heating or drying air.

Ingredient characteristics.	
Additive	*Fluid.*
Water or fat base, composition, concentration, viscosity, temperature, melting point, surface tension, setting behaviour	Nature, temperature, relative humidity

The combination of the above characteristics drives the choice of the process principle. It has then to be precisely described.

Process characteristics.	
Process	*Machine*
Continuous, batch, residence time, ingredient temperature, fluid temperature, system temperature, flow volumes, tolerance to variations, number of functions to fulfill (feeding, dosing, recycling, drying)	Form, internal surface, internal volume, size, mechanical movement, speed, temperature

The selection of the proper process and its control rely on the gathering of precise and reliable information.

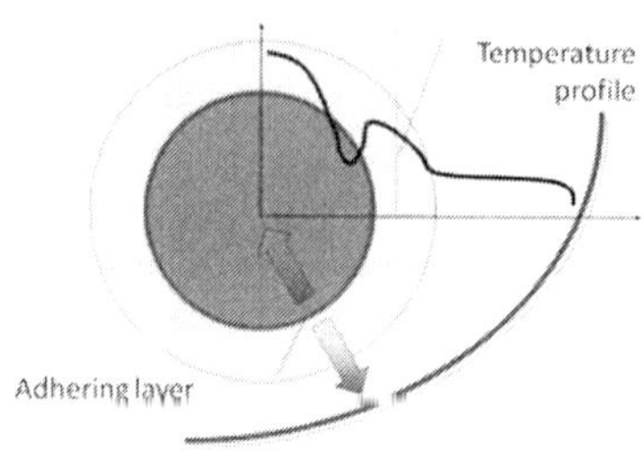

The influence of some phenomena and their parametres is critical: crystallisation, water removal (drying), glass transition, viscosity, or surface tension.

Temperature Among the parametres, temperature has a choice place. It influences viscosity, surface tension, drying or crystallisation behaviour. Ultimately, it influences the coating rate (thickness, weight gain) and coating resistance. It therefore influences the degree of clogging of product and ingredient in the system. For example, fat will tend to set preferably on a cool product if the system wall is kept at a higher temperature.

Practice

Coating Ingredients

Ingredient	*Form*	*Usual coating rate*	*Ingredients Characteristics related to coating*	*Coating aim*	*Examples*
Water	Pure liquid	1-3%		Wetting, adhesion, weight gain	Dust prevention, freezer burn prevention
Water	Saturated steam.	1-3%		Wetting, adhesion	Gluing of sugar on candies
Alcohol	Solution 70%	0,1%	Alcohol rate	Antimicrobial, preservative, texture enhancement	Preservation of packed pastries
Resin (shellac)	Alcoholic solution 30%	1%	Film forming	Surface aspect, flow, moisture barrier	Glazing of chocolate dragees
Wax (beeswax, carnauba, candelilla)	Solid form, melted, micronized suspension in oil	1%	Melting point	Surface aspect, barrier	Glazing of sugar dragees, fruit waxing
Sugars (saccharose, glucose, honey, polyols)	Solution 70-90%	1-100%	Crystallisation form, concentration, temperature	Palatability, surface aspect (glazed or frosty)	Sugar-coated ready-to-eat cereals
Natural hydrocolloid (gum arabic, xanthan, guar gelatine	Solution 20-40%	3%	Film forming, barrier	Mechanical or chemical protection, carrier of additives	Precoating of dragee with gum arabic and sugar prior to dragee pan coating
Starches (native or modified)	Colloidal solution 20-40%	3%	Characteristics depend on physical and chemical modifications, substitutes for more expensive ingredients (gum arabic, gelatine, titanium dioxide, etc.)	Mechanical or physical protection	Coating of fries before frying to reduce oil pick up
Flours	Powder or thick suspensions 20-40%	10-20%	Film forming, charge, viscosity, baking expansion	Thickness, crispiness	Coating of nuggets with batter prior to breading
Mineral or organic powders (talc, cellulose, potato flour, cellulose, starch)	Pure powder	1%	Moisture or fat absorption	Anticaking, drying	Coating of shredded cheese to prevent agglomeration
Oils and fats	Pure	1 - 40%	Melting point, viscosity	Anticaking, adhesion, barrier against moisture migration	Oiling of dry raisins or inclusions in ice cream
Seasonings, flavours, flavour enhancers, salt	Powder, diluted or concentrated solutions	1-3%	Concentration	Palatability	Flavouring of expanded snacks, salting of roasted nuts

Coating Techniques

For the sake of classification, two categories can be split easily into batch or continuous processes. Then, the categories can be refined according to the way the product is set in motion and the ingredient applied. Then, techniques allow either for just coating or can combine coating and setting in the same equipment.

Techniques				
Name	***Principle***	***Example***	***Alternative***	***Batch/Continuous***
Coextrusion	Forming of an outside casing around an inner content	Forming machine for sausages with forming of a collagen casing, further setting by coagulation and drying	High-temperature, short-time cooking-extrusion of snacks with continuous filling of a flavoured paste.	Continuous
Paddle mixer	Mixing by agitation in a closed volume	Snack coating	Helicoidal, scraping paddles	Batch
Vat mixer.	IQF coating (and freezing) tumbler.	Vacuum mixing		Batch
Conveyor	Application of the ingredient on the product spread across a conveyor.	Topical coating of pastries.	Spraying, screen, dipping coating.	Continuous.
Drum.	Application of the ingredient while the product is tumbled in drum.	Snack seasoning. Simultaneous coating and drying of cereals with sugar.		Continuous.
Screw.	Application of the ingredient while the product is transferred and mixed in a trough fitted with screw(s)).	Petfood.	Twin-screw systems for a better mixing.	Continuous.

Criteria for the Selection of a Technique.

- Base product characteristics : shape, size, bulk density. Size is the first criterion.
- Mechanical resistance of the base product.
- Final thickness of the coating layer.
- Complete/partial, top/side/bottom coating.
- Number of sequences to repeat.

- Processing time for each sequence.
- Setting mode : drying, cooling, freezing...
- Capacity.
- Preferred batch or continuous system.

Comparison batch vs. Continuous. The demand for higher yields makes production managers want to shift from manual batch to continuous automated systems. One has to consider the pros and contras prior to go for a costly and risky decision.

Comparison		
	Batch.	***Continuous.***
Advantages	Flexible, easy to monitor and control, quick response, less time-dependent, easy recipe change, full traceability, tool for R&D.	Efficient, justified if upstream and downstream processes are continuous.
Drawbacks.	Limited capacity. Manpower.	Expansive. Requires careful controls, feedback signals, consistent feeding and multiple peripherals.

Peripherals

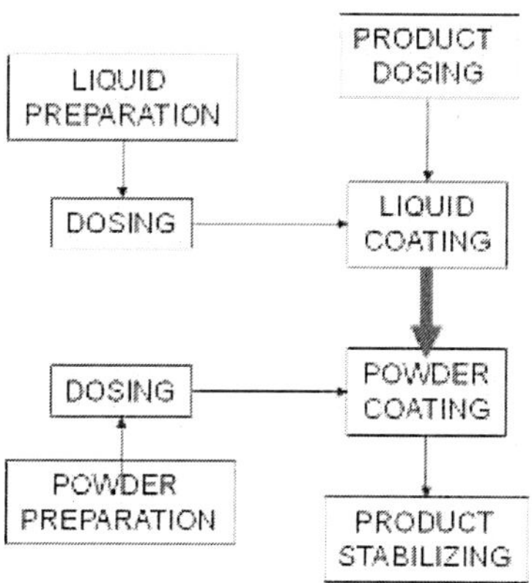

Given the number of operations and steps, a coating process can be an extensive process considered as a whole. The process core machine requires peripherals to serve it. A few frequent ones are listed for information.

- Storage.
- Ingredient preparation.
- Product feeding and metreing.
- Ingredient dosing.
- Filtration or sieving.
- Application system.
- Recycling.

Measures

Test results can be immediately evaluated (visual aspect) but are preferably assessed by careful measures : to allow monitoring, to agree on commissioning, to certify conformity with customer requirements.

Typical Measures

- Optical : colour, microscopy (homogeneity, thickness), image analysis.
- Weighing : weighing before and after treatment, weighing between batches or individual particles.
- Specific measurements according to target : compaction, barrier property.

Fat Migration

The migration of fat, moisture, and ethanol is a common problem with chocolate-coated confectionery products. Migration of one of these components into the coating leads to visual and sensory defects such as sugar or fat bloom, making the product unacceptable to the consumer. The migration rate depends on the structure and composition of the coating. The migration of each of these species can be slowed to a certain extent by proper tempering of the coating, because proper tempering will give a structure that resists migration. In the continuous lipid phase, these chemical species migrate mainly through the liquid portion. Thus, the migration rate depends on the amount of liquid oil present in the product. Migration can be delayed either by reducing the liquid fat content or by immobilizing the liquid phase. The actual mechanisms for the migration processes are speculative, and a more thorough understanding is necessary to better abate quality deterioration. Armed with this understanding, a manufacturer would know a priori the effect of changing the ingredient or process. A few methods for control have been suggested, but have found limited application. Mathematical models have been proposed to predict the migration behaviour, but their application is hindered because of the simplified assumptions employed. There is a need for developing better models that combine mass transfer with the phase behaviour to be able to accurately predict the migration process.

Moisture and Alcohol Migration

The migration of fat, moisture, and ethanol is a common problem with chocolate-coated confectionery products. Migration of one of these components into the coating leads to visual and sensory defects such as sugar or fat bloom, making the product unacceptable to the consumer. The migration rate depends on the structure and composition of the coating. The migration of each of these species can be slowed to a certain extent by proper tempering of the coating, because proper tempering will give a structure that resists migration. In the continuous lipid phase, these chemical species migrate mainly through the liquid portion. Thus, the migration rate depends on the amount of liquid oil present in the product. Migration can be delayed either by reducing the liquid fat content or by immobilizing the liquid phase. The actual mechanisms for the migration processes are speculative, and a more thorough understanding is necessary to better abate quality deterioration.

Rancidity

Rancidification, the product of which can be described as rancidity, is the process which causes a substance to become rancid, that is, having a rank, unpleasant smell or taste. Specifically, it is the hydrolysis and/or autoxidation of fats into short-chain aldehydes and ketones which are objectionable in taste and odour. When these processes occur in food, undesirable odors and flavours can result. In some cases, however, the flavours can be desirable (as in aged cheeses). In processed meats, these flavours are collectively known as warmed-over flavour. Rancidification can also detract from the nutritional value of food, and some vitamins are highly sensitive to degradation. Akin to rancidification, oxidative degradation also occurs in other hydrocarbons, e.g. lubricating oils, fuels, and mechanical cutting fluids.

Rancidification Pathways

Three pathways for rancidification are recognized:

Hydrolytic Rancidity

Hydrolytic rancidity refers to the odour that develops when triglycerides are hydrolyzed and free fatty acids are released. This reaction of lipid with water sometimes requires a catalyst, but results in the formation of free fatty acids and salts of free fatty acids. In particular, short-chain fatty acids, such as common butter fats, are odorous. Rancidity in foods may be very slight, indicated by a loss of freshness to very severe, indicated by objectionable odors and/or flavours. Slight degrees of rancidity are much more common in foods than severe rancidity, yet slight rancidity is a much more practical concern. A slight degree of rancidity may not be objectionable to consumers, but products which do not seem fresh will not attract repeat purchases. If customers do not return to a product, the long-term effects of a slight degree of rancidity can be economically serious.

Oxidative Rancidity

Oxidative rancidity is associated with the degradation by oxygen in the air. Via a free radical process, the double bonds of an unsaturated fatty acid can undergo cleavage, releasing volatile aldehydes and ketones. Oxidation primarily occurs with unsaturated fats. For example, even though meat is held under refrigeration or in a frozen state, the poly-unsaturated fat will continue to oxidize and slowly become rancid. The fat oxidation process, potentially resulting in rancidity, begins immediately after the animal is slaughtered and the muscle, intra-muscular, inter-muscular and surface fat becomes exposed to oxygen of the air. This chemical process continues during frozen storage, though more slowly at lower temperature. The process can be suppressed by the exclusion of oxygen or by the addition of antioxidants. Thus, airtight packaging will slow rancidity development.

Microbial Rancidity

Microbial rancidity refers to a process in which microorganisms, such as bacteria or molds, use their enzymes such as lipases to break down fat. This pathway can be prevented by sterilization.

Health Effects

Consuming rancid food products is unlikely to cause immediate illness or harm. Rancidification can reduce the nutritional value of food, and some vitamins are highly sensitive to degradation. In addition, rancidification can produce potentially toxic compounds associated with long-term harmful health effects concerning advanced aging, neurological disorders, heart disease, and cancer.

Reducing Rancidification

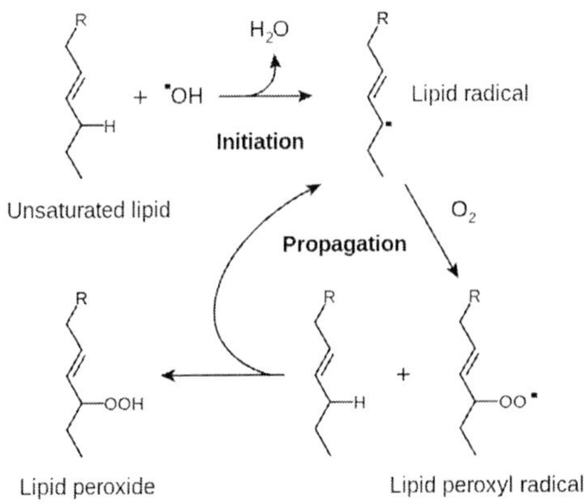

Figure: *The free radical pathway for the first phase of the oxidative rancidification of fats.*

Antioxidants are often used as preservatives in fat-containing foods to delay the onset or slow the development of rancidity due to oxidation. Natural antioxidants include polyphenols (for instance flavonoids), ascorbic acid (vitamin C) and tocopherols (vitamin E). Synthetic antioxidants include butylated hydroxyanisole (BHA), butylated hydroxytoluene (BHT), TBHQ, propyl gallate and ethoxyquin. The natural antioxidants tend to be short-lived, so synthetic antioxidants are used when a longer shelf-life is preferred.

The effectiveness of water-soluble antioxidants is limited in preventing direct oxidation within fats, but is valuable in intercepting free radicals that travel through the aqueous parts of foods. A combination of water-soluble and fat-soluble antioxidants is ideal, usually in the ratio of fat to water.

In addition, rancidification can be decreased, but not completely eliminated, by storing fats and oils in a cool, dark place with little exposure to oxygen or free radicals, since heat and light accelerate the rate of reaction of fats with oxygen. Antimicrobial agents can also delay or prevent rancidification by inhibiting the growth of bacteria or other micro-organisms that effect the process.

Oxygen scavenging technology can be used to remove oxygen from food packaging and therefore prevent oxidative rancidification.

Measurement of Oxidative Stability

Oxidative stability is a measure of an oil or fat's resistance to oxidation. Because the process takes place through a chain reaction, the oxidation reaction has a period when it is relatively slow, before it suddenly speeds up. The time for this to happen is called the "induction time", and it is repeatable under identical conditions (temperature, air flow, etc.). There are a number of ways to measure the progress of the oxidation reaction. One of the most popular methods currently in use is the Rancimat method.

The Rancimat method is carried out using an air current at temperatures between 50 and 220 °C. The volatile oxidation products (largely formic acid) are carried by the air current into the measuring vessel, where they are absorbed (dissolve) in the measuring fluid (distilled water).

By continuous measurement of the conductivity of this solution, oxidation curves can be generated. The cusp point of the oxidation curve (the point where a rapid rise in the conductivity starts) gives the induction time of the rancidification reaction, and can be taken as an indication of the oxidative stability of the sample.

The Rancimat method, the oxidative stability instrument (OSI) and the oxidograph were all developed as automatic versions of the more complicated AOM (active oxygen method), which is based on measuring peroxide values, for determining the induction time of fats and oils. Over time, the Rancimat method has become established, and it has been accepted into a number of national and international standards, for example AOCS Cd 12b-92 and ISO 6886.

Nutritional Aspects of Confectionery Fats

Confectionery (lollies and sweets) is a food of minimal nutritional value that is highly processed, high in sugar and often high in fat as well. Because of this, even small serves contribute a significant amount to energy intake while providing little nutritional benefit. The added sugar in confectionary can be in various forms such as sucrose, glucose, fructose, lactose, dextrose, maltose, maltodextrin, glucose syrups, fruit sugar, syrups, sugar-derived products such as molasses or treacle and icing sugar. Confectionery can also be made with artificial sweeteners such as sorbitol, mannitol, maltitol, xylitol, isomalt and polydextrose.

Chapter 6

Spreadable Products

Spreadable fats undergo standard classification according to their fat content and whether they are of dairy/non-dairy origin, vegetable origin or animal origin.

As of 1 July 2008, products falling within the scope of this Regulation are governed by the common organisation of agricultural markets.

This Regulation protects consumers from the possibility of confusing butter, margarine and other spreadable fats (e.g. minarines), by differentiating them according to their percentage of fat content and their animal or vegetable origin.

Definition

Spreadable fats are products with a fat content of at least 10% but less than 90% by weight and which remain solid at a temperature of 20°C.

Scope

To avoid any possible confusion, the Regulation limits the use of the terms "butter" and "margarine" to products with a fat content of not less than 80%.

"Reduced fat" Claims

Under the terms of the Regulation, the fact that the product has a reduced fat content must be mentioned clearly in the product designation. The Regulation therefore permits the use of nutritional claims which underline that the product has a reduced fat content. (Such claims consist of labelling, presentation and advertising information informing consumers about the characteristics of a food or of one of its ingredients.)

Sales and Import Descriptions

The various sales descriptions which are permitted, such as "minarine", "butter" and "cream" or the terms "vegetable" or "traditional" are defined in Articles 3 and 4.

Spreadable fats which are imported from non-Community countries are subject to the same requirements as those manufactured in the European Union (EU).

Butter

Butter is a dairy product made by churning fresh or fermented cream or milk, to separate the butterfat from the buttermilk. It is generally used as a spread and a condiment, as well as in cooking, such as baking, sauce making, and pan frying. Butter consists of butterfat, milk proteins and water.

Most frequently made from cows' milk, butter can also be manufactured from the milk of other mammals, including sheep, goats, buffalo, and yaks. Salt, flavourings and preservatives are sometimes added to butter. Rendering butter produces clarified butter or *ghee*, which is almost entirely butterfat.

Butter is a water-in-oil emulsion resulting from an inversion of the cream, an oil-in-water emulsion; the milk proteins are the emulsifiers. Butter remains a solid when refrigerated, but softens to a spreadable consistency at room temperature, and melts to a thin liquid consistency at 32–35 °C (90–95 °F). The density of butter is 911 g/L (56.9 lb/ft^3). It generally has a pale yellow colour, but varies from deep yellow to nearly white. Its unmodified colour is dependent on the animals' feed and is commonly manipulated with food colourings in the commercial manufacturing process, most commonly annatto or carotene.

Production

Unhomogenized milk and cream contain butterfat in microscopic globules. These globules are surrounded by membranes made of phospholipids (fatty acid emulsifiers) and proteins, which prevent the fat in milk from pooling together into a single mass. Butter is produced by agitating cream, which damages these membranes and allows the milk fats to conjoin, separating from the other parts of the cream. Variations in the production method will create butters with different consistencies, mostly due to the butterfat composition in the finished product. Butter contains fat in three separate forms: free butterfat, butterfat crystals, and undamaged fat globules. In the finished product, different proportions of these forms result in different consistencies within the butter; butters with many crystals are harder than butters dominated by free fats.

Churning produces small butter grains floating in the water-based portion of the cream. This watery liquid is called buttermilk—although the buttermilk most common today is instead a directly fermented skimmed milk. The buttermilk is drained off; sometimes more buttermilk is removed by rinsing the grains with water. Then the grains are "worked": pressed and kneaded together. When prepared manually, this is done using wooden boards called scotch hands. This consolidates the butter into a solid mass and breaks up embedded pockets of buttermilk or water into tiny droplets.

Commercial butter is about 80% butterfat and 15% water; traditionally made butter may have as little as 65% fat and 30% water. Butterfat is a mixture of triglyceride, a triester derived from glycerol and three of any of several fatty acid groups. Butter becomes rancid when these chains break down into smaller components, like butyric acid and diacetyl. The density of butter is 0.911 g/cm^3 (0.527 oz/in^3), about the same as ice.

Types

Before modern factory butter making, cream was usually collected from several milkings and was therefore several days old and somewhat fermented by the time it was made into butter. Butter made from a fermented cream is known as cultured butter. During fermentation, the cream naturally sours as bacteria convert milk sugars into lactic acid. The fermentation process produces additional aroma compounds, including diacetyl, which makes for a fuller-flavoured and more "buttery" tasting product. Today, cultured butter is usually made from pasteurized cream whose fermentation is produced by the introduction of Lactococcus and Leuconostoc bacteria.

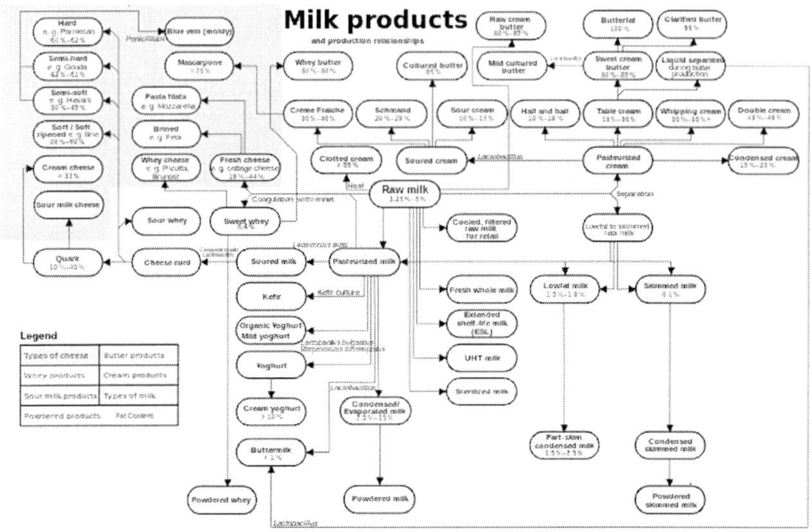

Figure: *Chart of milk products and production relationships, including butter.*

Another method for producing cultured butter, developed in the early 1970s, is to produce butter from fresh cream and then incorporate bacterial cultures and lactic acid. Using this method, the cultured butter flavour grows as the butter is aged in cold storage. For manufacturers, this method is more efficient, since aging the cream used to make butter takes significantly more space than simply storing the finished butter product. A method to make an artificial simulation of cultured butter is to add lactic acid and flavour compounds directly to the fresh-cream butter; while this more efficient process is claimed to simulate the taste of cultured butter, the product produced is not cultured but is instead flavoured.

Dairy products are often pasteurized during production to kill pathogenic bacteria and other microbes. Butter made from pasteurized fresh cream is called sweet cream butter. Production of sweet cream butter first became common in the 19th century, with the development of refrigeration and the mechanical cream separator. Butter made from fresh or cultured unpasteurized cream is called raw cream butter. While butter made from pasteurized cream may keep for several months, raw cream butter has a shelf life of roughly ten days.

Throughout continental Europe, cultured butter is preferred, while sweet cream butter dominates in the United States and the United Kingdom. Therefore, cultured butter is sometimes labelled "European-style" butter in the United States. Commercial raw cream butter is virtually unheard-of in the United States. Raw cream butter is generally only found made at home by consumers who have purchased raw whole milk directly from dairy farmers, skimmed the cream themselves, and made butter with it. It is rare in Europe as well.

Several "spreadable" butters have been developed; these remain softer at colder temperatures and are therefore easier to use directly out of refrigeration. Some modify the makeup of the butter's fat through chemical manipulation of the finished product, some through manipulation of the cattle's feed, and some by incorporating vegetable oils into the butter. "Whipped" butter, another product designed to be more spreadable, is aerated via the incorporation of nitrogen gas—normal air is not used, because doing so would encourage oxidation and rancidity.

All categories of butter are sold in both salted and unsalted forms. Either granular salt or a strong brine are added to salted butter during processing. In addition to enhanced flavour, the addition of salt acts as a preservative.

The amount of butterfat in the finished product is a vital aspect of production. In the United States, products sold as "butter" are required to contain a minimum of 80% butterfat; in practice, most American butters contain only slightly more than that, averaging around 81% butterfat. European butters generally have a higher ratio, which may extend up to 85%.

***Figure:** Liquid clarified butter*

Clarified butter is butter with almost all of its water and milk solids removed, leaving almost-pure butterfat. Clarified butter is made by heating butter to its melting point and then allowing it to cool; after settling, the remaining components separate by density. At the top, whey proteins form a skin which is removed, and the resulting butterfat is then poured off from the mixture of water and casein proteins that settle to the bottom.

Ghee is clarified butter which is brought to higher temperatures of around 120 °C (250 °F) once the water had evaporated, allowing the milk solids to brown. This process flavours the ghee, and also produces antioxidants which help protect it longer from rancidity. Because of this, ghee can keep for six to eight months under normal conditions.

Whey Butter

Cream may be separated (usually by a centrifugal separator) from whey instead of milk, as a byproduct of cheese-making. Whey butter may be made from whey cream. Whey cream and butter have a lower fat content and taste more salty, tangy and "cheesy". They are also cheaper than "sweet" cream and butter. The fat content of whey is low, so 1000 pounds of whey will typically give 3 pounds of butter.

European Butters

There are several butters produced in Europe with Protected geographical indications; these include:

- Beurre d'Ardenne, from Belgium
- Beurre d'Isigny, from France
- Beurre Charentes-Poitou (Which also includes: Beurre des Charentes and Beurre des Deux-Sèvres under the same classification), from France
- Beurre Rose, from Luxembourg
- Mantequilla de Soria, from Spain
- Mantega de l'Alt Urgell i la Cerdanya, from Spain

Size and Shape of Butter Packaging

Butter has traditionally been made into small, rectangular blocks by means of a pair of wooden butter paddles. In the United States, butter is usually produced in 4-ounce sticks, wrapped in waxed or foiled paper and sold four to a one-pound carton. This practice is believed to have originated in 1907, when Swift and Company began packaging butter in this manner for mass distribution.

Due to historical differences in butter printers (the machines which cut and package butter), these sticks are commonly produced in two different shapes:

- The dominant shape east of the Rocky Mountains is the Elgin, or Eastern-pack shape, named for a dairy in Elgin, Illinois. The sticks are 121 millimetres (4.8 in) long and 32 millimetres (1.3 in) wide and are typically sold stacked two by two in elongated cube-shaped boxes.
- West of the Rocky Mountains, butter printers standardized on a different shape that is now referred to as the Western-pack shape. These butter sticks are 80 millimetres (3.1 in) long and 38 millimetres (1.5 in) wide and are usually sold with four sticks packed side-by-side in a flat, rectangular box.

Both sticks contain the same amount of butter, although most butter dishes are designed for Elgin-style butter sticks.

The stick's wrapper is usually marked off as eight tablespoons (120 ml or 4.2 imp fl oz or 4.1 US fl oz); the actual volume of one stick is approximately nine tablespoons (130 ml or 4.6 imp fl oz or 4.4 US fl oz).

Outside of the United States, butter is packaged and sold by weight only, not by volume (fluid measure) nor by unit (stick), but the package shape remains approximately the same. The wrapper is usually a foil and waxed-paper laminate (the waxed paper is now a siliconised substitute, but is still referred to in some places as parchment, from the wrapping used in past centuries; and the term 'parchment-wrapped' is still employed where the paper alone is used, without the foil laminate).

In the UK and Ireland, and in some other regions historically accustomed to using British measures, this was traditionally ½lb and 1 lb packs; since metrication, pack sizes have changed to similar metric sizes such as 250g and 500g. In cooking (recipes), butter is specified and measured by weight only (grams or ounces); although melted butter could be measured by fluid measure (centilitres or fluid ounces), this is rare.

In the remainder of the metricated world, butter is packed and sold in 250g and 500g packs (roughly equivalent to the ½lb and 1 lb measures) and measured for cooking in grams or kilograms.

Butter for commercial and industrial use is packaged in plastic buckets, tubs, or drums, in quantities and units suited to the local market.

Worldwide

In 1997, India produced 1,470,000 metric tons (1,620,000 short tons) of butter, most of which was consumed domestically. Second in production was the United States (522,000 t or 575,000 short tons), followed by France (466,000 t or 514,000 short tons), Germany (442,000 t or 487,000 short tons), and New Zealand (307,000 t or 338,000 short tons). France ranks first in per capita butter consumption with 8 kg per capita per year.

In terms of absolute consumption, Germany was second after India, using 578,000 metric tons (637,000 short tons) of butter in 1997, followed by France (528,000 t or 582,000 short tons), Russia (514,000 t or 567,000 short tons), and the United States (505,000 t or 557,000 short tons).

New Zealand, Australia, and the Ukraine are among the few nations that export a significant percentage of the butter they produce.

Different varieties are found around the world. Smen is a spiced Moroccan clarified butter, buried in the ground and aged for months or years. Yak butter is a speciality in Tibet; tsampa, barley flour mixed with yak butter, is a staple food.

Butter tea is consumed in the Himalayan regions of Tibet, Bhutan, Nepal and India. It consists of tea served with intensely flavoured—or "rancid"—yak butter and salt. In African and Asian developing nations, butter is traditionally made from sour milk rather than cream. It can take several hours of churning to produce workable butter grains from fermented milk.

Storage and Cooking

Figure: *Hollandaise sauce served over white asparagus and potatoes.*

Normal butter softens to a spreadable consistency around 15 °C (60 °F), well above refrigerator temperatures. The "butter compartment" found in many refrigerators may be one of the warmer sections inside, but it still leaves butter quite hard. Until recently, many refrigerators sold in New Zealand featured a "butter conditioner", a compartment kept warmer than the rest of the refrigerator—but still cooler than room temperature—with a small heater. Keeping butter tightly wrapped delays rancidity, which is hastened by exposure to light or air, and also helps prevent it from picking up other odors. Wrapped butter has a shelf life of several months at refrigerator temperatures. "French butter dishes" or "Acadian butter dishes" involve a lid with a long interior lip, which sits in a container holding a small amount of water.

Usually the dish holds just enough water to submerge the interior lip when the dish is closed. Butter is packed into the lid. The water acts as a seal to keep the butter fresh, and also keeps the butter from overheating in hot temperatures.

This allows butter to be safely stored on the countertop for several days without spoilage.

Once butter is softened, spices, herbs, or other flavouring agents can be mixed into it, producing what is called a *compound butter* or *composite butter* (sometimes also called *composed butter*). Compound butters can be used as spreads, or cooled, sliced, and placed onto hot food to melt into a sauce. Sweetened compound butters can be served with desserts; such hard sauces are often flavoured with spirits.

Figure: *When heated, butter quickly melts into a thin liquid.*

Melted butter plays an important role in the preparation of sauces, most obviously in French cuisine. *Beurre noisette* (hazelnut butter) and *Beurre noir* (black butter) are sauces of melted butter cooked until the milk solids and sugars have turned golden or dark brown; they are often finished with an addition of vinegar or lemon juice. Hollandaise and béarnaise sauces are emulsions of egg yolk and melted butter; they are in essence mayonnaises made with butter instead of oil. Hollandaise and béarnaise sauces are stabilized with the powerful emulsifiers in the egg yolks, but butter itself contains enough emulsifiers—mostly remnants of the fat globule membranes—to form a stable emulsion on its own.

Beurre blanc (white butter) is made by whisking butter into reduced vinegar or wine, forming an emulsion with the texture of thick cream.

Beurre monté (prepared butter) is melted but still emulsified butter; it lends its name to the practice of "mounting" a sauce with butter: whisking cold butter into any water-based sauce at the end of cooking, giving the sauce a thicker body and a glossy shine—as well as a buttery taste.

In Poland, the butter lamb (*Baranek wielkanocny*) is a traditional addition to the Easter Meal for many Polish Catholics. Butter is shaped into a lamb either by hand or in a lamb-shaped mould. Butter is also used to make edible decorations to garnish other dishes.

Figure: *Mixing melted butter with chocolate to make a brownie*

Butter is used for sautéing and frying, although its milk solids brown and burn above 150 °C (250 °F)—a rather low temperature for most applications.

The smoke point of butterfat is around 200 °C (400 °F), so clarified butter or ghee is better suited to frying. Ghee has always been a common frying medium in India, where many avoid other animal fats for cultural or religious reasons.

Butter fills several roles in baking, where it is used in a similar manner as other solid fats like lard, suet, or shortening, but has a flavour that may better complement sweet baked goods.

Many cookie doughs and some cake batters are leavened, at least in part, by creaming butter and sugar together, which introduces air bubbles into the butter.

The tiny bubbles locked within the butter expand in the heat of baking and aerate the cookie or cake. Some cookies like shortbread may have no other source of moisture but the water in the butter. Pastries like pie dough incorporate pieces of solid fat into the dough, which become flat layers of fat when the dough is rolled out. During baking, the fat melts away, leaving a flaky texture.

Butter, because of its flavour, is a common choice for the fat in such a dough, but it can be more difficult to work with than shortening because of its low melting point.

Pastry makers often chill all their ingredients and utensils while working with a butter dough.

Stabilisers

A stabiliser is an additive to food which helps to preserve its structure. Typical uses include preventing oil/water emulsions from separating in products such as salad dressing; preventing ice crystals from forming in frozen food such as ice cream; and preventing fruit from settling in products such as jam and yogurt.

The following hydrocolloids are the most common ones used as stabilisers:

- alginate
- agar
- carrageen
- cellulose and cellulose derivatives
- gelatin
- guar gum
- gum Arabic
- locust bean gum
- pectin
- starch
- xanthan gum

Fatty Acid

A fatty acid is a carboxylic acid with a long aliphatic tail (chain), which is either saturated or unsaturated. Most naturally occurring fatty acids have a chain of an even number of carbon atoms, from 4 to 28. Fatty acids are usually derived from triglycerides or phospholipids. When they are not attached to other molecules, they are known as "free" fatty acids.

Fatty acids are important sources of fuel because, when metabolized, they yield large quantities of ATP. Many cell types can use either glucose or fatty acids for this purpose.

In particular, heart and skeletal muscle prefer fatty acids. Despite long-standing assertions to the contrary, fatty acids can be used as a source of fuel for brain cells, at least in some rodents, in addition to glucose and ketone bodies.

Types of Fatty Acids

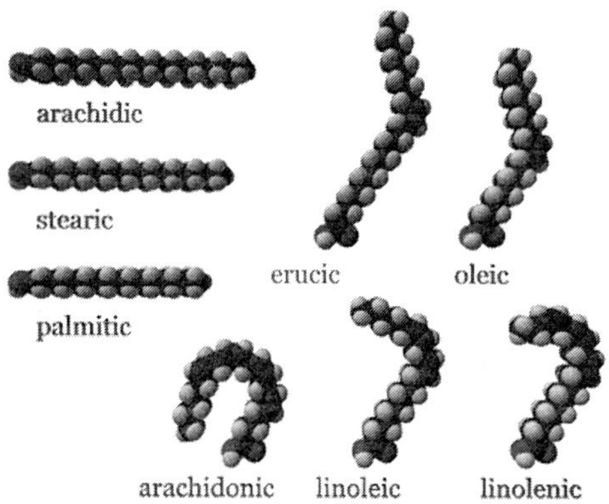

Figure: *Three-dimensional representations of several fatty acids*

Fatty acids that have carbon–carbon double bonds are known as unsaturated. Fatty acids without double bonds are known as saturated. They differ in length as well.

Length of Free Fatty acid Chains

Fatty acid chains differ by length, often categorized as short to very long.

- Short-chain fatty acids (SCFA) are fatty acids with aliphatic tails of fewer than six carbons (i.e. butyric acid).
- Medium-chain fatty acids (MCFA) are fatty acids with aliphatic tails of 6–12 carbons, which can form medium-chain triglycerides.
- Long-chain fatty acids (LCFA) are fatty acids with aliphatic tails 13 to 21 carbons.
- Very long chain fatty acids (VLCFA) are fatty acids with aliphatic tails longer than 22 carbons.

Unsaturated Fatty Acids

Unsaturated fatty acids have one or more double bonds between carbon atoms. (Pairs of carbon atoms connected by double bonds can be saturated by adding hydrogen atoms to them, converting the double bonds to single bonds. Therefore, the double bonds are called unsaturated.)

The two carbon atoms in the chain that are bound next to either side of the double bond can occur in a *cis* or *trans* configuration.

cis

A cis configuration means that the two hydrogen atoms adjacent to the double bond stick out on the same side of the chain. The rigidity of the double bond freezes its conformation and, in the case of the cis isomer, causes the chain to bend and restricts the conformational freedom of the fatty acid. The more double bonds the chain has in the cis configuration, the less flexibility it has. When a chain has many cis bonds, it becomes quite curved in its most accessible conformations. For example, oleic acid, with one double bond, has a "kink" in it, whereas linoleic acid, with two double bonds, has a more pronounced bend. Alpha-linolenic acid, with three double bonds, favours a hooked shape. The effect of this is that, in restricted environments, such as when fatty acids are part of a phospholipid in a lipid bilayer, or triglycerides in lipid droplets, cis bonds limit the ability of fatty acids to be closely packed, and therefore can affect the melting temperature of the membrane or of the fat.

Trans

A trans configuration, by contrast, means that the adjacent two hydrogen atoms lie on opposite sides of the chain. As a result, they do not cause the chain to bend much, and their shape is similar to straight saturated fatty acids.

In most naturally occurring unsaturated fatty acids, each double bond has three n carbon atoms after it, for some n, and all are cis bonds. Most fatty acids in the trans configuration (trans fats) are not found in nature and are the result of human processing (e.g., hydrogenation).

The differences in geometry between the various types of unsaturated fatty acids, as well as between saturated and unsaturated fatty acids, play an important role in biological processes, and in the construction of biological structures (such as cell membranes).

Examples of Unsaturated Fatty Acids				
Common name	***Chemical structure***	Δ^x	***C:D***	***n–x***
Myristoleic acid	$CH_3(CH_2)_3CH=CH(CH_2)_7COOH$	*cis*-Δ^9	14:1	*N*–5
Palmitoleic acid	$CH_3(CH_2)_5CH=CH(CH_2)_7COOH$	*cis*-Δ^9	16:1	*N*–7
Sapienic acid	$CH_3(CH_2)_8CH=CH(CH_2)_4COOH$	*cis*-Δ^6	16:1	*N*–10
Oleic acid	$CH_3(CH_2)_7CH=CH(CH_2)_7COOH$	*cis*-Δ^9	18:1	*n*–9
Elaidic acid	$CH_3(CH_2)_7CH=CH(CH_2)_7COOH$	*trans*-Δ^9	18:1	*n*–9
Vaccenic acid	$CH_3(CH_2)_5CH=CH(CH_2)_9COOH$	*trans*-Δ^{11}	18:1	*N*–7
Linoleic acid	$CH_3(CH_2)_4CH=CHCH_2CH=CH(CH_2)_7COOH$	*cis,cis*-Δ^9,Δ^{12}	18:2	*n*–6
Linoelaidic acid	$CH_3(CH_2)_4CH=CHCH_2CH=CH(CH_2)_7COOH$	*trans,trans*-Δ^9,Δ^{12}	18:2	*n*–6
α-Linolenic acid	$CH_3CH_2CH=CHCH_2CH=CHCH_2CH=CH(CH_2)_7COOH$	*cis,cis,cis*-$\Delta^9,\Delta^{12},\Delta^{15}$	18:3	*n*–3
Arachidonic acid	$CH_3(CH_2)_4CH=CHCH_2CH=CHCH_2CH=CHCH_2CH=CH(CH_2)_3COOH$NIST	*cis,cis,cis,cis*-$\Delta^5\Delta^8,\Delta^{11},\Delta^{14}$	20:4	*n*?6
Eicosapentaenoic acid	$CH_3CH_2CH=CHCH_2CH=CHCH_2CH=CHCH_2CH=CHCH_2CH=CH(CH_2)_3COOH$	*cis,cis,cis,cis,cis*-$\Delta^5,\Delta^8,\Delta^{11},\Delta^{14},\Delta^{17}$	20:5	*n*–3
Erucic acid	$CH_3(CH_2)_7CH=CH(CH_2)_{11}COOH$	*cis*-Δ^{13}	22:1	*n*–9
Docosahexaenoic acid	$CH_3CH_2CH=CHCH_2CH=CHCH_2CH=CHCH_2CH=CHCH_2CH=CHCH_2CH=CH(CH_2)_2COOH$	*cis,cis,cis,cis,cis,cis*-$\Delta^4,\Delta^7,\Delta^{10},\Delta^{13},\Delta^{16},\Delta^{19}$	22:6	*n*–3

Essential Fatty Acids

Fatty acids that are required by the human body but cannot be made in sufficient quantity from other substrates, and therefore must be obtained from food, are called essential fatty acids. There are two series of essential fatty acids: one has a double bond three carbon atoms removed from the methyl end; the other has a double bond six carbon atoms removed from the methyl end. Humans lack the ability to introduce double bonds in fatty acids beyond carbons 9 and 10, as counted from the carboxylic acid side. Two essential fatty acids are linoleic acid (LA) and alpha-linolenic acid (ALA). They are widely distributed in plant oils. The human body has a limited ability to convert ALA into the longer-chain *n*-3 fatty acids eicosapentaenoic acid (EPA) and docosahexaenoic acid (DHA), which can also be obtained from fish.

Saturated Fatty Acids

Saturated fatty acids are long-chain carboxylic acids that usually have between 12 and 24 carbon atoms and have no double bonds.

Thus, saturated fatty acids are saturated with hydrogen (since double bonds reduce the number of hydrogens on each carbon). Because saturated fatty acids have only single bonds, each carbon atom within the chain has 2 hydrogen atoms (except for the omega carbon at the end that has 3 hydrogens).

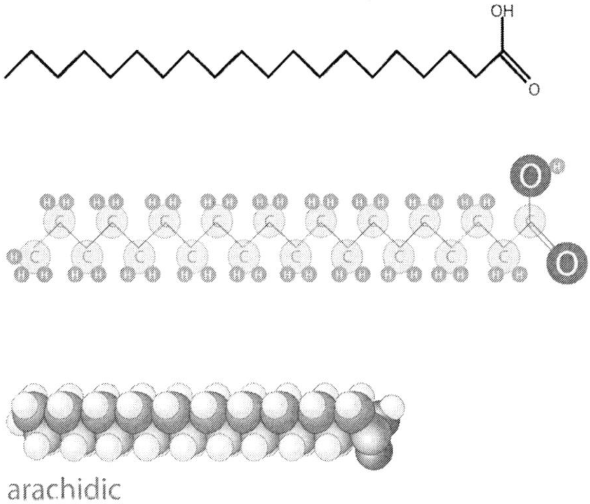

Figure: *Arachidic acid, a saturated fatty acid.*

Examples of Saturated Fatty Acids

Common name	***Chemical structure***	***C:D***
Caprylic acid	$CH_3(CH_2)_6COOH$	8:0
Capric acid	$CH_3(CH_2)_8COOH$	10:0
Lauric acid	$CH_3(CH_2)_{10}COOH$	12:0
Myristic acid	$CH_3(CH_2)_{12}COOH$	14:0
Palmitic acid	$CH_3(CH_2)_{14}COOH$	16:0
Stearic acid	$CH_3(CH_2)_{16}COOH$	18:0
Arachidic acid	$CH_3(CH_2)_{18}COOH$	20:0
Behenic acid	$CH_3(CH_2)_{20}COOH$	22:0
Lignoceric acid	$CH_3(CH_2)_{22}COOH$	24:0
Cerotic acid	$CH_3(CH_2)_{24}COOH$	26:0

Nomenclature

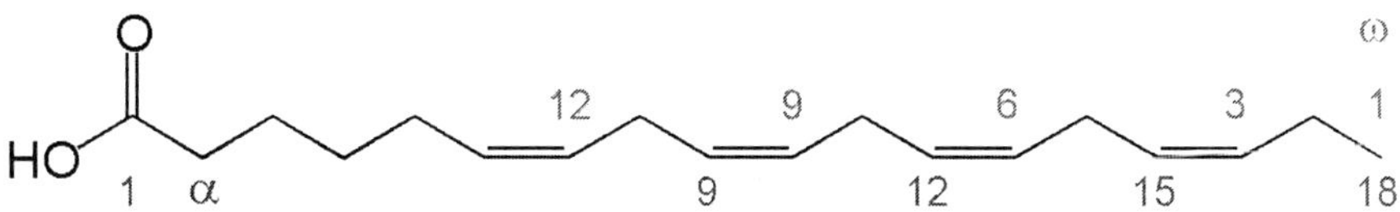

Figure: *Numbering of carbon atoms*

Several different systems of nomenclature are used for fatty acids. The following table describes the most common systems.

System	*Example*	*Explanation*
Trivial nomenclature	Palmitoleic acid	Trivial names (or common names) are non-systematic historical names, which are the most frequent naming system used in literature. Most common fatty acids have trivial names in addition to their *systematic names* (see below). These names frequently do not follow any pattern, but they are concise and often unambiguous.
Systematic nomenclature	(9Z)-octadecenoic acid	Systematic names (or IUPAC names) derive from the standard *IUPAC Rules for the Nomenclature of Organic Chemistry*, published in 1979,[8] along with a recommendation published specifically for lipids in 1977.[9] Counting begins from the carboxylic acid end. Double bonds are labelled with *cis-/trans-* notation or *E-/Z-* notation, where appropriate. This notation is generally more verbose than common nomenclature, but has the advantage of being more technically clear and descriptive.
Δ^x nomenclature	*cis,cis*-Δ^9,Δ^{12} octadecadienoic acid	In Δ^x (or delta-*x*) nomenclature, each double bond is indicated by Δ^x, where the double bond is located on the *x*th carbon–carbon bond, counting from the carboxylic acid end. Each double bond is preceded by a *cis-* or *trans-* prefix, indicating the conformation of the molecule around the bond. For example, linoleic acid is designated "cis-Δ^9, cis-Δ^{12} octadecadienoic acid". This nomenclature has the advantage of being less verbose than systematic nomenclature, but is no more technically clear or descriptive.
n? *x* nomenclature	*n*? 3	*n*? *x* (*n* minus *x*; also ω? *x* or omega-*x*) nomenclature both provides names for individual compounds and classifies them by their likely biosynthetic properties in animals. A double bond is located on the xth carbon–carbon bond, counting from the terminal methyl carbon (designated as *n* or ω) toward the carbonyl carbon. For example, α-Linolenic acid is classified as a *n*? 3 or omega-3 fatty acid, and so it is likely to share a biosynthetic pathway with other compounds of this type. The ω? *x*, omega-*x*, or "omega" notation is common in popular nutritional literature, but IUPAC has deprecated it in favor of *n*? *x* notation in technical documents. The most commonly researched fatty acid biosynthetic pathways are *n*? 3 and *n*? 6.
Lipid numbers	18:3 18:3ω6 18:3, cis,cis,cis-Δ^9,Δ^{12},Δ^{15}	Lipid numbers take the form *C*:*D*, where *C* is the number of carbon atoms in the fatty acid and *D* is the number of double bonds in the fatty acid (if more than one, the double bonds are assumed to be interrupted by CH 2 units, *i.e.*, at intervals of 3 carbon atoms along the chain). This notation can be ambiguous, as some different fatty acids can have the same numbers. Consequently, when ambiguity exists this notation is usually paired with either a Δ^x or *n*? *x* term.

Production

Fatty acids are usually produced industrially by the hydrolysis of triglycerides, with the removal of glycerol. Phospholipids represent another source. Some fatty acids are produced synthetically by hydrocarboxylation of alkenes.

Free Fatty Acids

The biosynthesis of fatty acids involves the condensation of acetyl-CoA. Since this coenzyme carries a two-carbon-atom group, almost all natural fatty acids have even numbers of carbon atoms.

The "uncombined fatty acids" or "free fatty acids" found in organisms come from the breakdown of a triglyceride. Because they are insoluble in water, these fatty acids are transported (solubilized, circulated) while bound to plasma protein albumin. The levels of "free fatty acid" in the blood are limited by the availability of albumin binding sites.

Fatty Acids in Dietary Fats

The following table gives the fatty acid, vitamin E and cholesterol composition of some common dietary fats.

	Saturated	*Monounsaturated*	*Polyunsaturated*	*Cholesterol*	*Vitamin E*
	g/100g	g/100g	g/100g	mg/100g	mg/100g
Animal fats					
Lard	40.8	43.8	9.6	93	0.60
Duck fat	33.2	49.3	12.9	100	2.70
Butter	54.0	19.8	2.6	230	2.00
Vegetable fats					
Coconut oil	85.2	6.6	1.7	0	.66
Palm kernel oil	81.5	11.4	1.6	0	3.80
Palm oil	45.3	41.6	8.3	0	33.12
Cottonseed oil	25.5	21.3	48.1	0	42.77
Wheat germ oil	18.8	15.9	60.7	0	136.65
Soybean oil	14.5	23.2	56.5	0	16.29
Olive oil	14.0	69.7	11.2	0	5.10
Corn oil	12.7	24.7	57.8	0	17.24
Sunflower oil	11.9	20.2	63.0	0	49.00
Safflower oil	10.2	12.6	72.1	0	40.68
Hemp oil	10	15	75	0	12.34
Canola/Rapeseed oil	5.3	64.3	24.8	0	22.21

Reactions of Fatty Acids

Fatty acids exhibit reactions like other carboxylic acids, i.e. they undergo esterification and acid-base reactions.

Acidity

Fatty acids do not show a great variation in their acidities, as indicated by their respective pK_a. Nonanoic acid, for example, has a pK_a of 4.96, being only slightly weaker than acetic acid (4.76). As the chain length increases, the solubility of the fatty acids in water decreases very rapidly, so that the longer-chain fatty acids have minimal effect on the pH of an aqueous solution. Even those fatty acids that are insoluble in water will dissolve in warm ethanol, and can be titrated with sodium hydroxide solution using phenolphthalein as an indicator to a pale-pink endpoint. This analysis is used to determine the free fatty acid content of fats; i.e., the proportion of the triglycerides that have been hydrolyzed.

Hydrogenation and Hardening

Hydrogenation of unsaturated fatty acids is widely practiced to give saturated fatty acids, which are less prone towards rancidification.

Since the saturated fatty acids are higher melting than the unsaturated relatives, the process is called hardening. This technology is used to convert vegetable oils into margarine. During partial hydrogenation, unsaturated fatty acids can be isomerized from *cis* to *trans* configuration.

More forcing hydrogenation, i.e. using higher pressures of H_2 and higher temperatures, converts fatty acids into fatty alcohols. Fatty alcohols are, however, more easily produced from fatty acid esters.

In the Varrentrapp reaction certain unsaturated fatty acids are cleaved in molten alkali, a reaction at one time of relevance to structure elucidation.

Auto-oxidation and Rancidity

Unsaturated fatty acids undergo a chemical change known as auto-oxidation. The process requires oxygen (air) and is accelerated by the presence of trace metals. Vegetable oils resist this process because they contain antioxidants, such as tocopherol. Fats and oils often are treated with chelating agents such as citric acid to remove the metal catalysts.

Ozonolysis

Unsaturated fatty acids are susceptible to degradation by ozone. This reaction is practiced in the production of azelaic acid ($(CH_2)_7(CO_2H)_2$) from oleic acid.

Analysis

In chemical analysis, fatty acids are separated by gas chromatography of methyl esters; additionally, a separation of unsaturated isomers is possible by argentation thin-layer chromatography.

Yolk Fat

The yolk is a part of an egg (or just of the egg cell in non-egg-laying animals) that feeds the developing embryo in animals. In whole eggs, 43% of the protein comes from the yolk.

Also known as deutoplasm, this food material is accumulated during oogenesis, together with RNA molecules and other substances. These may be synthesized by the oocyte itself or by the cytoplasm of other non-germ cells. The amount of yolk in an egg cell affects the developmental processes that follow fertilization.

Apart from animals, other organisms, like algae, specially in the oogamous, can also accumulate resources in their female gametes. In gymnosperms, the remains of the female gametophyte serve also as food supply, and in flowering plants, the endosperm.

Chicken Egg Yolk

In the avian egg, the yolk usually is some shade of yellow in colour. It is spherical and is suspended in the egg white (known alternatively as albumen or glair/glaire) by one or two spiral bands of tissue called the chalazae. Prior to fertilization, the yolk is a single cell, the ovum or egg cell, one of the few single cells that can be seen by the naked eye. This fact was discovered by Hoyer in 1858.

After the fertilization, the cleavage leads to the formation of the germinal disc. As a food, chicken egg yolks are a major source of vitamins and minerals. They contain all of the egg's fat and cholesterol, and about one-half of the protein. If left intact while cooking fried eggs, the yellow yolk surrounded by a flat blob of whites creates a distinctive sunny-side up form. Mixing the two components together before frying results in a pale yellow mass, as in omelettes and scrambled eggs.

Uses

- It is at times separated from the egg white and used in cooking (for mayonnaise, custard, hollandaise sauce, *crème brûlée*, *avgolemono*, and *ovos moles*).
- It is used in painting as a component of traditional egg-tempera.
- It is used in the production of egg-yolk agar plate medium, useful in testing for the presence of *Clostridium perfringens*.
- Egg yolk contains an antibody called antiglobulin (IgY). The antibody transfers from the laying hen to the egg yolk by passive immunity to protect both embryo and hatchling from microorganism invasion.
- Egg yolk can be used to make liqueurs such as Advocaat or eggnog.
- Egg yolk is used to extract egg oil which has various cosmetic, nutritional and medicinal uses.
- The developing embryo inside the egg uses the yolk as sustenance.

Composition of Chicken Egg Yolk

The yolk makes up about 33% of the liquid weight of the egg; it contains approximately 60 calories, three times the caloric content of the egg white.

The yolk of one large egg (50 g total, 17 g yolk) contains approximately: 2.7 g protein, 210 mg cholesterol, 0.61 g carbohydrates, and 4.51 g total fat.

All of the fat-soluble vitamins (A, D, E, and K) are found in the egg yolk. Egg yolk is one of the few foods naturally containing vitamin D.

The composition (by weight) of the most prevalent fatty acids in egg yolk is typically as follows:

- Unsaturated fatty acids:
    - Oleic acid, 47%
    - Linoleic acid, 16%
    - Palmitoleic acid, 5%
    - Linolenic acid, 2%
- Saturated fatty acids:
    - Palmitic acid, 23%
    - Stearic acid, 4%
    - Myristic acid, 1%

Egg yolk is a source of lecithin as well as egg oil for cosmetic and pharmaceutical applications. Based on weight, egg yolk contains about 9% lecithin.

The yellow colour is due to lutein and zeaxanthin, which are yellow or orange carotenoids known as xanthophylls.

Yolk Proteins

The different yolk's protein has distinct roles. Phosvitins are important in sequestering calcium, iron and other cations for the developing embryo. Phosvitins are one of the most phosphorylated (10%) proteins in nature, the high concentration of phosphate groups providing efficient metal-binding sites in clusters. Lipovitellins are involved in lipid and metal storage, and contain a heterogeneous mixture of about 16% (w/w) noncovalently bound lipid, most being phospholipid. Lipovitellin-1 contains two chains, LV1N and LV1C.

Yolk Vitamins and Minerals

Yolks hold more than 90% of the calcium, iron, phosphorus, zinc, thiamine, B6, folate, and B12, and pantothenic acid of the egg. In addition, yolks cover all of the fat soluble vitamins A, D, E, and K in the egg, as well as all of the essential fatty acids (EFAs).

A single yolk from a large egg contains approximately 22 mg of calcium, 66 mg of phosphorus, 9.5 micrograms of selenium and 19 mg of potassium, according to the USDA. Refer to the chart at right for quantities of nutrients per 100 g of egg yolk.

Double-yolk Eggs

Double-yolk eggs occur when ovulation occurs too rapidly, or when one yolk becomes joined with another yolk. These eggs may be the result of a young hen's reproductive cycle not yet being synchronized.

Double-yolked eggs seldom lead to successful hatchlings without human intervention, as the chicks interfere with each other's hatching process and do not survive.

Higher-order yolks are rare, although heavier poultry breeds such as the Buff Orpington have been known to lay triple-yolk eggs occasionally.

Yolkless Eggs

Eggs without yolks are known as "dwarf" or "wind" eggs, or the archaic term "cock egg". Such an egg is most often a pullet's first effort, produced before her laying mechanism is fully ready. Mature hens rarely lay a yolkless egg, but it sometimes happens that a piece of reproductive tissue breaks away and passes down the tube. Such a scrap of tissue may stimulate the egg-producing glands to react as though it were a yolk and wrap it in albumen, membranes and a shell as it travels through the egg tube. This is usually what causes an egg to contain a small particle of grayish tissue instead of a yolk.

Since these eggs contain no yolk and therefore can not hatch, they were traditionally believed to have been laid by roosters. This type of egg occurs in many varieties of fowl and has been found in chickens, both standard and bantams, guineas and coturnix quail.

Yolk Colour

The colour of an egg yolk is directly influenced by the makeup of the chicken feed. Egg yolk colour is generally improved with a feed containing a large component of yellow, fat-soluble pigments, such as the carotenes in dark green plant material, for example alfalfa. Although much emphasis is put onto the colour of the egg yolk, it does not reliably reflect the nutritional value of an egg. For example, some of the natural pigments that produce a rich yolk colour are xanthophylls without much nutritional value, rather than the carotenoids that act as provitamin A in the body. Also, a diet rich in vitamin A itself, but without A-provitamins or xanthophylls, can produce practically colourless yolks that are just as nutritious as any richly coloured yolks. Since unhealthy chickens produce fewer and smaller eggs, farmers ensure that whatever the source of their feed, the quality is adequate, so there is not likely to be much difference in the nutritional quality of the eggs.

Yolks, particularly from free-range eggs, can be of a wide range of colours, ranging from nearly white, through yellow and orange to practically red, but even olive green, depending on the pigments in their food. Feeding fowls large amounts of capsicum peppers for example, tends to result in red or deep orange yolks. This has nothing to do with adding colours such as cochineal to eggs in cooking.

Monoglycerides

Monoglyceride is a glyceride in which each glycerol molecule has formed an ester bond with exactly one fatty acid molecule. The more formally correct terms in modern convention are acylglycerol and monoacylglycerol.

Any monoacylglycerol is either a 1-monoacylglycerol or a 2-monoacylglycerol, depending on the position of the ester bond on the glycerol moiety. 1-monoacylglycerides possess a chiral centre at carbon 2.

Monoacylglycerols are variously produced by biological or industrial chemical processes. Biochemically they commonly are formed by enzymatic hydrolysis of triglycerides by the action of lipoprotein lipase, by enzymatic hydrolysis of a fatty acid from diacylglycerol by the action of diacylglycerol lipase, or as an intermediate in the alkanoylation of glycerol to form fats.

In normal metabolic processes monoacylglycerols are hydrolysed by monoacylglycerol lipase to produce glycerol and a free fatty acid as required.

Together with diglycerides, monoglycerides commonly are added to commercial food products in small quantities. In these applications they are useful as emulsifiers, helping to mix ingredients such as oily materials and water that otherwise would blend poorly. The values given in the nutritional labels for total fat, saturated fat, and *trans* fat do not include those present in mono- and diglycerides, largely because the bulk of the fat content consists of triacylglycerols.

The commercial raw materials for the production of monoacylglycerols may be either vegetable or animal fats and oils. Bulk animal sources typically are cattle or hogs, but any economically available triacylglycerols with acceptable fatty-acid composition may be used as preferred. Monoacylglycerols also may be made synthetically. They are often found in bakery products, beverages, ice cream, chewing gum, shortening, whipped toppings, margarine, and confections. In bakery products, monoglycerides are useful in improving loaf volume and texture, and as antistaling agents.

One particular monoacylglycerol, 2-arachidonoylglycerol, is a full agonist of CB_1 cannabinoid receptors. Another monoacylglycerol agonist is 2-oleoylglycerol, which is a GPR119 agonist.

Lecithin

Lecithin is a generic term to designate any group of yellow-brownish fatty substances occurring in animal and plant tissues composed of phosphoric acid, choline, fatty acids, glycerol, glycolipids, triglycerides, and phospholipids (e.g., phosphatidylcholine, phosphatidylethanolamine, and phosphatidylinositol).

Lecithin was first isolated in 1846 by the French chemist and pharmacist Theodore Gobley. In 1850, he named the phosphatidylcholine *léchithine*. Gobley originally isolated lecithin from egg yolk—λÝêéèïò *lckithos* is "egg yolk" in Ancient Greek—and established the complete chemical formula of phosphatidylcholine in 1874; in between, he had demonstrated the presence of lecithin in a variety of biological matters, including venous blood, bile, human brain tissue, fish eggs, fish roe, and chicken and sheep brain.

Lecithin can easily be extracted chemically (using hexane, ethanol, acetone, petroleum ether, benzene, etc.) or mechanically. It is usually available from sources such as soybeans, eggs, milk, marine sources, rapeseed, cottonseed, and sunflower. It has low solubility in water, but is an excellent emulsifier. In aqueous solution, its phospholipids can form either liposomes, bilayer sheets, micelles, or lamellar structures, depending on hydration and temperature. This results in a type of surfactant that usually is classified as amphipathic. Lecithin is sold as a food supplement and for medical uses. In cooking, it is sometimes used as an emulsifier and to prevent sticking, for example in nonstick cooking spray.

Lecithin, as a food additive, is also a dietary source of several active compounds: Choline and its metabolites are needed for several physiological purposes, including cell membrane signaling and cholinergic neurotransmission, and is a major source for methyl groups via its metabolite, trimethylglycine (betaine). Phosphatidylcholine occurs in all cellular organisms, being one of the major components of the phospholipid portion of the cell membrane.

While lecitihin is also a rich source of a variety of types of dietary fats, the small amounts of lecithin typically used for food additive purposes mean it is not a significant source of fats.

Production

Commercial lecithin, as used by food manufacturers, is a mixture of phospholipids in oil. The lecithin can be obtained by water degumming the extracted oil of seeds. It is a mixture of various phospholipids, and the composition depends on the origin of the lecithin. A major source of lecithin is soybean oil. Because of the EU requirement to declare additions of allergens in foods, in addition to regulations regarding genetically modified crops, a gradual shift to other sources of lecithin (e.g., sunflower oil) is taking place. The main phospholipids in lecithin from soya and sunflower are phosphatidyl choline, phosphatidyl inositol, phosphatidyl ethanolamine, and phosphatidic acid. They often are abbreviated to PC, PI, PE, and PA, respectively. Purified phospholipids are produced by companies commercially.

Hydrolysed Lecithin

To modify the performance of lecithin to make it suitable for the product to which it is added, it may be hydrolysed enzymatically. In hydrolysed lecithins, a portion of the phospholipids have one fatty acid removed by phospholipase. Such phospholipids are called lysophospholipids. The most commonly used phospholipase is phospholipase A2, which removes the fatty acid at the C2 position of glycerol. Lecithins may also be modified by a process called fractionation. During this process, lecithin is mixed with an alcohol, usually ethanol. Some phospholipids, such as phosphatidylcholine, have good solubility in ethanol, whereas most other phospholipids do not dissolve well in ethanol. The ethanol is separated from the lecithin sludge, after which the ethanol is removed by evaporation to obtain a phosphatidylcholine-enriched lecithin fraction.

Genetically Modified Crops as a Source of Lecithin

As described above, lecithin is highly processed. Therefore, genetically modified (GM) protein or DNA from the original GM crop from which it is derived often is undetectable – in other words, it is not substantially different from lecithin derived from non-GM crops. Nonetheless, consumer concerns about genetically modified food have extended to highly purified derivatives from GM food, such as lecithin. This concern led to policy and regulatory changes in the European Union in 2000, when Commission Regulation (EC) 50/2000 was passed which required labelling of food containing additives derived from GMOs, including lecithin. Because it is nearly impossible to detect the origin of derivatives such as lecithin, the European regulations require those who wish to sell lecithin in Europe to use a meticulous system of identity preservation (IP).

Properties and Applications

Lecithin has emulsification and lubricant properties, and is a surfactant. It can be totally metabolized by humans, so is well tolerated by humans and nontoxic when ingested; some other emulsifiers can only be excreted via the kidneys.

The major components of commercial soybean-derived lecithin are:

- 33–35% Soybean oil
- 20–21% Inositol phosphatides
- 19–21% Phosphatidylcholine
- 8–20% Phosphatidylethanolamine
- 5–11% Other phosphatides
- 5% Free carbohydrates
- 2–5% Sterols
- 1% Moisture

Lecithin is used for applications in human food, animal feed, pharmaceuticals, paints, and other industrial applications.

Applications Include

- In the pharmaceutical industry, it acts as a wetting, stabilizing agent and a choline enrichment carrier, helps in emulsifications and encapsulation, and is a good dispersing agent. It can be used in manufacture of intravenous fat infusions and for therapeutic use.
- In animal feed, it enriches fat and protein and improves pelletization.
- In the paint industry, it forms protective coatings for surfaces with painting and printing ink, has antioxidant properties, helps as a rust inhibitor, is a colour-intensifying agent, catalyst, conditioning aid modifier, and dispersing aid; it is a good stabilizing and suspending agent, emulsifier, and wetting agent, helps in maintaining uniform mixture of several pigments, helps in grinding of metal oxide pigments, is a spreading and

mixing aid, prevents hard settling of pigments, eliminates foam in water-based paints, and helps in fast dispersion of latex-based paints.

- Lecithin also may be used as a release agent for plastics, an antisludge additive in motor lubricants, an antigumming agent in gasoline, and an emulsifier, spreading agent, and antioxidant in textile, rubber, and other industries.

Food Additive

The nontoxicity of lecithin leads to its use with food, as an additive or in food preparation. It is used commercially in foods requiring a natural emulsifier or lubricant.

In confectionery, it reduces viscosity, replaces more expensive ingredients, controls sugar crystallization and the flow properties of chocolate, helps in the homogeneous mixing of ingredients, improves shelf life for some products, and can be used as a coating. In emulsions and fat spreads, it stabilizes emulsions, reduces spattering during frying, improves texture of spreads and flavour release. In doughs and bakery, it reduces fat and egg requirements, helps even distribution of ingredients in dough, stabilizes fermentation, increases volume, protects yeast cells in dough when frozen, and acts as a releasing agent to prevent sticking and simplify cleaning.

It improves wetting properties of hydrophilic powders (e.g., low-fat proteins) and lipophilic powders (e.g., cocoa powder), controls dust, and helps complete dispersion in water. Lecithin keeps cocoa and cocoa butter in a candy bar from separating. It can be used as a component of cooking sprays to prevent sticking and as a releasing agent. In margarines, especially those containing high levels of fat (>75%), lecithin is added as an 'antispattering' agent for shallow frying.

Lecithin is approved by the United States Food and Drug Administration for human consumption with the status "generally recognized as safe". Lecithin is admitted by the EU as a food additive, designated as E322. Research studies show soy-derived lecithin has significant effects on lowering serum cholesterol and triglycerides, while increasing HDL ("good cholesterol") levels in the blood of rats.

Dietary Supplement

Because it contains phosphatidylcholines, lecithin is a source of choline, an essential nutrient. Clinical studies have shown benefit in acne, in improving liver function, and in lowering cholesterol, but clinical studies in dementia and dyskinesias have found no benefit. An earlier study using a small sample (20 men divided in 3 groups) did not detect statistically significant short term (2-4 weeks) effects on cholesterol in hyperlipidaemic men.

La Leche League recommends its use to prevent blocked or plugged milk ducts which can lead to mastitis in breastfeeding women.

Compatibility with Special Diets

Egg-derived lecithin is not usually a concern for those allergic to eggs since commercially available egg lecithin is highly purified and devoid of allergy-causing egg proteins. Egg lecithin is not a concern for those on low-cholesterol diets, because the lecithin found in eggs markedly inhibits the absorption of the cholesterol contained in eggs.

Ghee

Ghee is a class of clarified butter that originated in ancient India and is commonly used in Kurdish, Afghani, Pakistani, Indian, Bangladeshi, Nepali and Sri Lankan cuisine, traditional medicine and religious rituals.

For centuries, clarified butter has had enormous cultural influence around the globe. Making clarified butter involves slowly heating *unsalted* butter to 212°F (100°C) and letting it boil until the water vaporizes. Once the bubbling stops, three layers remain: whey protein, liquid fat, and casein particles.

Once you remove the skin of whey protein, you can pour off the fat, which is the clarified butter.

For smaller quantities of clarified butter, use a heavy saucepan and watch for the telltale signs of clarification (white, and then brown specks at the bottom of the pan). Remove the pan from the heat, and set into a cold-water bath for a few seconds. Skim off the top layer, or strain the contents of the pan through a triple layer of cheesecloth or a coffee filter.

Clarified butter is perfect for frying over high heat, since the components that cause butter to scorch – protein and casein - have been removed. Regular butter can be used, though, for frying proteins over medium heat. In fact, the buttery flavour of the caramelized milk proteins will stick to the meat, and the fat can be poured off.

If you're frying breaded foods, make sure to use fine, fresh breadcrumbs and sauté using clarified butter. The result is extremely flavourful, but without the greasiness of breaded foods fried in oil.

Ghee

Ghee is the name for anhydrous butter fat in India, where it is prepared in large quantities; it is commonly mixed with the milk fat of the water-buffalo. The full name is *usli ghee*, and the spelling *ghi* is sometimes used, or in sanskrit *ghrta.*

Ghee is the chief form of cooking oil in many Indian regional cuisines; it is also used medicinally and plays a part in some Hindu religious ceremonies. [Read more on ghee in India.] *Samna* (also *samneh, samn*) is the name for butter fat in Egypt and elsewhere in the Middle East, where it is also prepared in large quantities; it is commonly mixed with the milk fats of sheep and goats. It turns up in North Africa as *sman*, sometimes flavoured with herbs, or spiced, or aged. Ethiopia also has a spiced version, *nit'r k'ibe.* [Read more on smen.]

The butter is melted and the simmered long enough to boil off all the water, during which time it takes on a buttery taste. It is used especially, but not exclusively, for cooking meat, and it is essential for many Indian dishes.

Ghee is the clear butter fat. By removing the albuminous curd and water that favour the growth of organisms promoting rancidity, anhydrous butter fat does not become rancid as readily as butter and can be stored unrefrigerated for several months.

Here is an Example Recipe: Ghee is produced as follows. Butter made from cow's milk is melted over a slow fire and then heated slowly until the separated water boils off. The vessel holding the butter is then allowed to cool; semifluid, clear butterfat, which makes the finest ghee, rises to the top of the melted butter and may be poured off, leaving the curd (precipitated protein) at the bottom of the vessel.

The curd, which still contains 50 percent or more butterfat, may be reworked with the addition of peanut oil or buffalo milk fat to make inferior grades of ghee.

A significant portion of Indian ghee is made from buffalo butter, but only ghee made from cow's butter has any religious or medical significance among Hindus. Early Sanskrit writings attributed many medicinal qualities to ghee, such as improving the voice and sight and increasing longevity. Ghee is used in almost every one of the numerous religious ceremonies that Hindus observe at different points in their lives, including birth, initiation into manhood, wedding sacrifices, and gift-giving at death. Images of the gods are washed in ghee, and it is frequently used to light holy lamps or is thrown upon an altar fire in sacrifice.

Making of Ghee Commercially

The processes for the production of anhydrous milkfat (butter oil), using cream as the raw material, are based on the emulsion splitting principle. In brief, the processes consist of the cream first being concentrated and the fat globules then being broken down mechanically, so that the fat is liberated. This forms a continuous fat phase containing dispersed water droplets which can be separated from the fat phase.

Either the Clarifixator or the Centrifixator is used for mechanically liberating the fat and thus allowing for the phase conversion, and these two process lines derive their names from the corresponding units.

The Clarifixator line has been used commercially for a number of years. One of the key machines in the system - the Clarifixator - is a centrifugal separator equipped with a serrated disc which simultaneously homogenizes the milk.

The disc normally breaks down the fat globules into smaller globules, although in the production of anhydrous milk fat, the disc breaks down the emulsion, so that the liquid leaving the machine is a continuous oil phase, with dispersed water droplets and buttermilk. After phase inversion, the fat is concentrated in a hermetic separator. Clarifixator lines with capacities between 500 and 1000 kg of butter oil per hour are available.

The Centrifixator line has an appreciably higher capacity of 1500 - 2000 kg of butter oil per hour and offers opportunities for planning production rates of 4000 - 6000 kg of butter oil per hour. In this case, the emulsion is split in a unit of special design - known as the Centrifixator - which contains a motor-driven serrated disc.

The fat is concentrated in a hermetic solids-ejecting separator. Solids-ejecting machines are also used in the second centrifugal separation stage, and the line is thus capable of operating for long periods and can be cleaned in place.

The market size of Ghee in India will double its size from US$ 60 Billion (INR 2.6 Trillion) to US$ 115 Billion (INR 5.1 Trillion) by 2016. India is the world's largest producer of Ghee and also its largest consumer.

Culinary Uses

Ghee is widely used in Indian cuisine. All over India, rice is sometimes traditionally prepared or served with ghee (including *biryani*). In Rajasthan, ghee is eaten with *baati*. All over north India, people sometimes dab *roti* with ghee.

In Bengal (both West Bengal and Bangladesh) and Gujarat, ghee is served with *kichdi*, which is an evening meal (or dinner) of rice with lentils cooked in curry made from yogurt, cumin seeds, curry leaves, ghee, cornflour, turmeric, garlic, and salt.

Figure: *A* dosa *in India served with ghee*

Ghee is also used to prepare *kadhi* and used in Indian sweets such as *Mysore pak*, and different varieties of *halva* and *laddu*. Punjabi cuisine prepared in restaurants uses large amounts of ghee. *Naan* and *roti* are sometimes brushed with ghee, either during preparation or while serving. Ghee is an important part of Punjabi cuisine and traditionally, the *parathas, daals*, and curries in Punjab often use ghee instead of oil, to make them rich in taste. Different types of ghees are used in different types of cooking recipes; for example, ghee made from cow's milk is traditionally served with rice or *roti* or just a generous sprinkle over the top of a curry or *daal* (lentils), but for cooking purposes, ghee made from buffalo's milk is used generally.

Ghee is an ideal fat for deep frying because its smoke point (where its molecules begin to break down) is 250 °C (482 °F), which is well above typical cooking temperatures of around 200 °C (392 °F) and above that of most vegetable oils.

Like any clarified butter, ghee is composed almost entirely of fat, 62% of which consists of saturated fats; the nutrition facts label found on bottled cow's ghee produced in the United States indicates 8 mg of cholesterol per teaspoon.

Indian restaurants and some households may use partially hydrogenated vegetable oil (also known as *vanaspati*, *dalda*, or "vegetable ghee") in place of ghee because of its lower cost. This vegetable ghee may contain trans fat. Trans fats have been shown to increase the risk of coronary heart disease even moreso than saturated fats.

The term *shuddh ghee*, however, is not used in many regions as partially hydrogenated oils are marketed as pure ghee in some areas. In India, the sale of fake ghee is stopped by law enforcement agencies whenever a complaint is made. Ghee is also sometimes called *desi* (country-made) ghee or *asli* (genuine) ghee to distinguish it from vegetable ghee.

Fats & fatty acids	***Amounts per 100 g of ghee***
Total fat	99.5 g (153% DV)
Saturated fat	61.9 g (310% DV)
Monounsaturated fat	28.7 g
Polyunsaturated fat	3.7 g
Trans fats	4 g
Omega-3 fatty acids	1447 mg
Omega-6 fatty acids	2247 mg

Other non-fat nutrients	***Amounts per 100 g of ghee***
Carbohydrates	0
Minerals	0
Cholesterol	256 mg (85%DV)
Phytosterols	0
Vitamin A	3069 IU (61% DV)
Vitamin B, C, D	0
Vitamin E	2.8 mg (14% DV)
Vitamin K	8.6 μg (11% DV)

Clarified Butter

Clarified butter is milk fat rendered from butter to separate the milk solids and water from the butterfat. Typically, it is produced by melting butter and allowing the components to separate by density. The water evaporates, some solids float to the surface and are skimmed off, and the remainder of the milk solids sink to the bottom and are left behind when the butter fat (which would then be on top) is poured off.

Commercial methods of production also include direct evaporation, but may also be accomplished by decantation and centrifugation followed by vacuum drying; or direct from cream by breaking the emulsion followed by centrifugation.

Figure: *Freshly made clarified butter, still liquid*

Figure: *Clarified butter at room temperature*

Properties

Clarified butter has a higher smoke point (485 °F or 252 °C) than regular butter (325-375 °F or 163-190 °C), and is therefore preferred in some cooking applications, such as sautéing. Clarified butter also has a much longer shelf life than fresh butter.

It has negligible amounts of lactose and casein and is, therefore, acceptable to most who have a lactose intolerance or milk allergy.

Regional Variations

In the Indian Subcontinent, South Asia (ghee), the butter may be cooked long enough to evaporate the water portion and caramelize the milk solids (which are then filtered out), resulting in a nutty flavour.

Vanaspati

Vanaspati, an alternative to ghee, is made from partially-hydrogenated permitted vegetable oils, and converted to solid form by hydrogenation. It is white and grainy (danedaar) in texture and is favoured for deep-frying. It is used as a ghee substitute in cooking as well in the production of bakery products, sweets and snack items.

When Vanaspati is made, trans-fatty acids are also produced; these increase the risk of heart disease when consumed in excess and are best avoided. Dalda is the best known brand of Vanaspati.

How to Select

- Choose a reliable brand and select a pack size as per your requirement.
- Sometimes, the product may be fortified with vitamin A & D for health benefits. So, read the labels carefully and choose.

Culinary Uses

- Vanaspati is so flavourful that it is enough to use just half or two-thirds as much as other cooking oils.
- It has a very high burning point and doesn't burn or smoke during cooking.
- It combines excellently with a wide variety of spices.
- You can use Vanaspati ghee to sauté foods, in baking, to deep-fry, or even as a spread.
- It is often used for frying, when making sweets like gulab jamun, jalebis etc.
- Vanaspati is very commonly used in preparing bakery foods like puffs, khari, breads and biscuits, nankhatai, cakes, sweets and ice-cream too.
- Although very economical, it is better to use Vanaspati ghee merely as a flavouring/seasoning rather than the primary fat due to the presence of trans-fatty acids.

How to Store

- Vanaspati ghee can be stored in an air-tight container. It does not need refrigeration.

Health Benefits

- Vanaspati ghee is most notably said to stimulate the secretion of stomach acids to help with digestion, while other fats, such as butter and oils, slow down the digestive process and can sit heavily in the stomach.

- It does not spoil easily, and actually preserves the original freshness and potency of herbs and foods.
- Helps balance excess stomach acid, and helps maintain/repair the mucus lining of the stomach.
- Vanaspati ghee is said to prevent blisters and scarring if applied quickly to affected skin.
- It is said to promote all three aspects of mental functioning — learning, memory and recall.
- Moderate intake proves beneficial; however excessive amounts can cause weight gain and other lifestyle diseases.

Speciality Oils

Almond Oil

Almonds are a rich source of oil, with values ranging between 36 to 60% of kernel dry mass. A study by Venkatchalam and Sathe suggests almonds contain approximately 44% oils, of which 62% is monounsaturated oleic acid (an omega-9 fatty acid), 29% is linoleic acid (a polyunsaturated omega-6 essential fatty acid), and 9% is saturated fatty acid.

"Oleum amygdalae", *the fixed oil, is prepared from either sweet or bitter almonds and is a glyceryl oleate, with a slight odour and a nutty taste. It is almost insoluble in alcohol but readily soluble in chloroform or ether. Sweet almond oil is obtained from the dried kernel of sweet almonds.*

The oil is good for application to the skin as an emollient, and has been traditionally used by massage therapists to lubricate the skin during a massage session.

Almond oil can also be used as a wood conditioner of certain woodwind instruments, such as the oboe and clarinet.

Groundnut Oil

Peanut oil, also known as groundnut oil or arachis oil, is a mild tasting vegetable oil derived from peanuts. The oil is available in refined, unrefined, cold pressed, and roasted varieties, the latter with a strong peanut flavour and aroma, analogous to sesame oil.

Filtered groundnut or peanut oil has been used in India since generations. Refining of oil was introduced in the 1950s; the first refined groundnut oil was called "Sipahi" but it gained popularity under the brand name "Postman".

Peanut oil is supposedly less refined than other oils in order to retain some of the peanut protein for taste and nutrition. Hence, peanut oil generally retains the aroma and flavour of peanuts to an extent.

It also has a high smoke point compared to other oils, and is hence used commonly for frying foods like French fries.

Its major component fatty acids are palmitic acid, oleic acid, and linoleic acid. The oil also contains six to eight per cent (total) of arachidic acid, arachidonic acid, behenic acid, lignoceric acid and other fatty acids.

It is often used in Chinese, South Asian and Southeast Asian cuisine, both for general cooking, and in the case of roasted oil, for added flavour. Peanut oil has a high smoke point relative to many other cooking oils, so is commonly used for frying foods. Its major component fatty acids are oleic acid (46.8% as olein), linoleic acid (33.4% as linolein), and palmitic acid (10.0% as palmitin). The oil also contains some stearic acid, arachidic acid, arachidonic acid, behenic acid, lignoceric acid and other fatty acids.

Antioxidants such as vitamin E are sometimes added, to improve the shelf life of the oil.

According to the USDA data upon which the following table is based, 100 g of peanut oil contains 17.7 g of saturated fat, 48.3 g of monounsaturated fat, and 33.4 g of polyunsaturated fat.

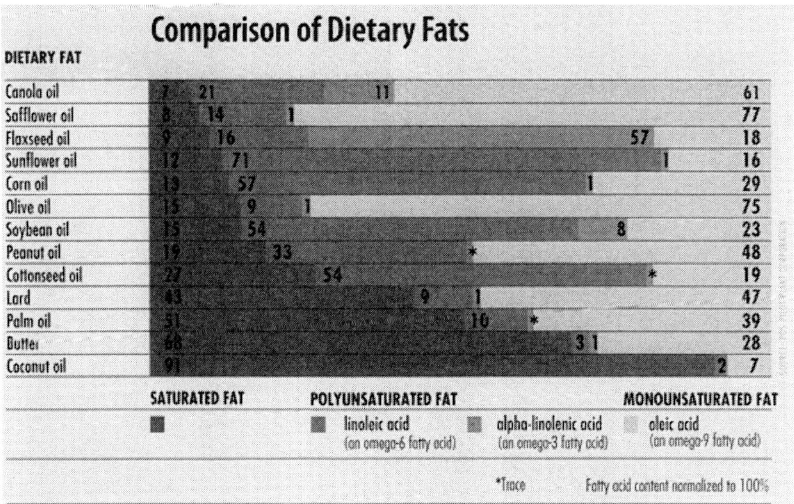

Hazelnut Oil

Hazelnut is the nut of the hazel and therefore includes any of the nuts deriving from species of the genus Corylus avellana. It is also known as cobnut or filbert nut according to species. A cob is roughly spherical to oval, about 15–25 millimetres (0.59–0.98 in) long and 10–15 millimetres (0.39–0.59 in) in diametre, with an outer fibrous husk surrounding a smooth shell. A filbert is more elongated, being about twice as long as it is round. The nut falls out of the husk when ripe, about seven to eight months after pollination. The kernel of the seed is edible and used raw or roasted, or ground into a paste. Hazelnuts are also used for livestock feed, as are chestnuts and acorns. The seed has a thin, dark brown skin, which is sometimes removed before cooking.

Hazelnut oil is a luscious, fragrant addition to dressings and baking. It is expensive and can oxidise quickly, so it's best to buy only as much as you need.

Hazelnut oil, as the name indicates, is made by pressing the oil from hazelnuts. It has an intense flavour—slightly sweet, roasted, and nutty. Though it's relatively expensive as oils go, you don't need to use much for the potent hazelnut flavour to come through.

Hazelnut oil is a popular addition to cookies, cakes, and other baked goods. It also adds rich, nutty flavour to sauces and salad dressings, but it's usually combined with more mildly-flavoured oils so as not to overpower the other flavours in the dish.

Like most nut oils, hazelnut oil needs to be stored in a cool place to keep it from going rancid.

Sesame Seed Oil

Sesame oil is an edible vegetable oil derived from sesame seeds. Besides being used as a cooking oil in South India, it is often used as a flavour enhancer in Chinese, Japanese, Middle Eastern, Korean, and Southeast Asian cuisine.

The oil from the nutrient-rich seed is popular in alternative medicine, from traditional massages and treatments to the modern day. The traditional Indian medical practice of Ayurveda uses sesame oil.

The oil is popular in Asia and is also one of the earliest-known crop-based oils, but world-wide mass modern production continues to be limited even today due to the inefficient manual harvesting process required to extract the oil.

Composition

Table: Sesame oil is composed of the following fatty acids:

Fatty acid	Nomenclature	Minimum	Maximum
Palmitic	C16:0	7.0%	12.0%
Palmitoleic	C16:1w7	trace	0.5%
Stearic	C18:0	3.5%	6.0%
Oleic	C18:1w9	35.0%	50.0%/lino
Linoleic	C18:2w6	35.0%	50.0%
Linolenic	C18:3w3	trace	1.0%
Eicosenoic	C20:1	trace	1.0%

Sesame oil:Specifications

Test	value
Density at 30°C / 30°C	0.915-0.919
Refractive Index(40°C)	1.4645-1.4665
Saponification value	188-193
Iodine value	103-120
unsaponifiyable matter % max	1.5-2.0%
Bellier turbidity temp.°C max	22°C

The lignans sesamin and sesamolin are minor components of sesame oil.

Manufacture of Sesame Oil

Manufacturing Process

Figure: Bottling sesame oil at Moran Market, Seongnam, Gyeonggi Province, South Korea.

Sesame seeds are protected by a capsule which only bursts when the seeds are completely ripe. The ripening time tends to vary, so farmers cut plants by hand and place them together in an upright position to continue ripening until all the capsules have opened. The discovery of an indehiscent (nonshattering) mutant by Langham in 1943 began the work towards development of a high yielding, shatter-resistant variety. Although researchers have made significant progress in sesame breeding, harvest losses due to shattering continue to limit domestic US production.

Sesame seeds are primarily produced in developing countries, a factor that has played a role in limiting the creation of large-scale, fully automated oil extraction and processing techniques. Sesame oil can be extracted by a number of methods, depending on the materials and equipment available.

In developing countries, sesame oil is often extracted with less-expensive and manually intensive techniques such as hot water flotation, bridge presses, ram presses, the ghani process, or by using a small-scale expeller. In developed countries sesame oil is often extracted using an expeller press, larger-scale oil extraction machines, or by pressing followed by chemical solvent extraction.

Sesame oil can also be extracted under low-temperature conditions using an expeller press in a process called cold pressing. This extraction method is popular among raw food adherents because it avoids exposing the oil to chemical solvents or high temperatures during extraction.

While some manufacturers will further refine sesame oil through solvent extraction, neutralization and bleaching in order to improve its cosmetic aspects, sesame oil derived from quality seeds already possesses a pleasant taste and does not require further purification before it can be consumed. Many consumers prefer unrefined sesame oil due to their belief that the refining process removes important nutrients.

Sesame oil is one of the more stable natural oils, but can still benefit from refrigeration and limited exposure to light and high temperatures during extraction, processing and storage in order to minimize nutrient loss through oxidation and rancidity. Storage in amber-coloured bottles can help to minimize light exposure.

Sesame oil is a polyunsaturated (PUFA) semi-drying oil. Commercial sesame oil varies in colour from light to deep reddish yellow depending on the colour of the seed processed and the method of milling. Provided the oil is milled from well-cleaned seed, it can be refined and bleached easily to yield a light-coloured limpid oil. Sesame oil is rich in oleic and linoleic acids, which together account for 85% of the total fatty acids. Sesame oil has a relatively high percentage of unsaponifiable matter (1.5-2.3%) in India and in some other European countries. It is obligatory to add sesame oil (5-10%) to margarine and generally to hydrogenated vegetable fats which are commonly used as adulterants for butter or ghee.

Sesame Seed Market

As of 2012, sesame seeds sold on the global market for roughly US $0.67/lb. This relatively high price reflects a worldwide shortage. Though the market for sesame seed is strong, domestic US production awaits the development of high-yielding nonshattering varieties. About 65 percent of the annual US sesame crop is processed into oil and 35 percent is used in food. The market for sesame oil is mainly located in Asia and the Middle East where the use of domestically produced sesame oil has been a tradition for centuries.

Varieties

There are many variations in the colour of sesame oil: cold-pressed sesame oil is pale yellow, while Indian sesame oil (gingelly or til oil) is golden, and East Asian sesame oils are commonly a dark brown colour. This dark colour and flavour are derived from roasted/ toasted sesame seeds. Cold pressed sesame oil has a different flavour than the toasted oil, since it is produced directly from raw, rather than toasted, seeds.

Sesame oil is traded in any of the forms described above: Cold-pressed sesame oil is available in Western health shops. Unroasted (but not necessarily cold pressed) sesame oil is commonly used for cooking in the Middle East and can often be found in halal markets. In East Asian countries, different kinds of hot-pressed sesame oil are preferred.

Uses

Cooking

Despite sesame oil's high proportion (41%) of polyunsaturated (Omega-6) fatty acids, it is least prone, among cooking oils with high smoke points, to turn rancid when kept in the open. This is due to the natural antioxidants present in the oil.

Light sesame oil has a high smoke point and is suitable for deep-frying, while dark sesame oil (from roasted sesame seeds) has a slightly lower smoke point and is unsuitable for deep-frying. Instead it can be used for the stir frying of meats or vegetables, sautéing, or for the making of an omelette. East Asian cuisines often use roasted sesame oil for seasoning.

The Chinese use sesame oil in the preparation of meals for women during postpartum confinement

Sesame oil is the traditional oil used for frying tempura in Japan, particularly in the Tokyo area.

Sesame oil is most popular in Asia, especially in Korea, China, and the South Indian states of Karnataka, Andhra Pradesh, and Tamil Nadu, where its widespread use is similar to that of olive oil in the Mediterranean.

In South India, sesame oil is used in almost all pickles and condiments. It is often mixed in with a special spice powder that accompanies Idly, dosa as well as rice mixed with spice powders ([Paruppu Podi]). It is used particularly in Tamil Nadu and Andhra Pradesh mixed in with foods that are hot and spicy as it neutralizes the heat. It is not used for frying as the oil is heavy by itself.

Traditional Uses in India

Sesame oil is reputed to penetrate the skin easily and is used in India for oil massage. Sesame oil (Til Tel) is specially used for massaging as it is believed to rid the body of heat due to its viscous nature upon rubbing. It is also used for hair and scalp massage. Sesame oil is used in the manufacture of Ayurvedic drugs. It is also used in many cosmetic applications, including as a carrier oil.

In Hinduism, sesame or "til" oil is used in deepa or oil lamps kept in front of shrines for the Deities. Sesame oil is used for performing puja in Hindu temples. Also, particularly in South India, sesame oil is applied to the stone deities in the temple's shrines. It is only used on Deities made of black granite.

Industrial Uses

In industry, sesame oil may be used as

- a solvent in injected drugs or intravenous drip solutions,
- a cosmetics carrier oil,
- coating stored grains to prevent weevil attacks. The oil also has synergy with some insecticides.

Low grade oil is used locally in soaps, paints, lubricants, and illuminants.

Nutrients

Sesame oil is a source of vitamin E. Sesame oil also contains magnesium, copper, calcium, iron, zinc, and vitamin B6.

Safflower Oil

*Safflower (*Carthamus tinctorius *L.) is a highly branched, herbaceous, thistle-like annual plant. It is commercially cultivated for vegetable oil extracted from the seeds. Plants are 30 to 150 cm (12 to 59 in) tall with globular flower heads having yellow, orange, or red flowers. Each branch will usually have from one to five flower heads containing 15 to 20 seeds per head. Safflower is native to arid environments having seasonal rain. It grows a deep taproot which enables it to thrive in such environments.*

There are two types of safflower that produce different kinds of oil: one high in monounsaturated fatty acid (oleic acid) and the other high in polyunsaturated fatty acid (linoleic acid). Currently the predominant edible oil market is for the former, which is lower in saturates than olive oil, for example. The latter is used in painting in the place of linseed oil, particularly with white paints, as it does not have the yellow tint which linseed oil possesses.

In dietary use, high–linoleic safflower oil has also been shown to increase adiponectin, a protein that helps regulate blood glucose levels and fatty-acid breakdown. During a 16-week, double-blind controlled study conducted at The Ohio State University, researchers compared high-linoleic safflower oil (SAF) with conjugated linoleic acid (CLA). They studied post-menopausal women who had high blood sugar and wanted to lose weight. These participants showed an average reduction of 6.3 percent belly fat and an average of 20.3 percent increase in the important belly fat hormone, adiponectin.

Hornstra et al analyzed a group where safflower oil replaced animal fats in the diets of people who had had a heart attack. The group receiving extra safflower oil in place of animal fats had a significantly higher risk of death from all causes, cardiovascular disease and coronary heart disease. As expected, increasing omega-6 linoleic acid from safflower oil in the Sydney Diet Heart Study significantly reduced total cholesterol; however, these reductions were not associated with [reduced] mortality outcomes. Moreover, the increased risk of death in the intervention group presented fairly rapidly and persisted throughout the trial. An updated meta-analysis of polyunsaturated fatty acid intervention trials showed trends towards increased risks of death from coronary heart disease and cardiovascular disease from increasing omega-6 linoleic acid intakes suggesting that the cardiovascular benefits of polyunsaturated fatty acids may be attributable to omega-3 polyunsaturated fats.

In culinary use, safflower oil compares favourably with other vegetable oils with its high smoke point.

Grape Seed Oil

Grape seed oil (also called grapeseed oil or grape oil) is pressed from the seeds of grapes, and is thus an abundant by-product of winemaking.

Uses

Cooking

Grape seed oil has a moderately high smoke point of approximately 216 °C (421 °F). Due to its clean, light taste, and high polyunsaturated fat content, it may be used as an ingredient in salad dressings and mayonnaise and as a base for oil infusions of garlic, rosemary, or other herbs or spices. It is widely used in baked goods, pancakes, and waffles. It is sprayed on raisins to help them retain their flavour.

The metabolic energy density of grape seed oil is typical of vegetable oils: approximately 3,700 kJ (880 kcal) per 100 g, or 500 kJ (120 kcal) per 15 ml tablespoon.

Cosmetics

Grape seed oil is a preferred cosmetic ingredient for controlling moisture of the skin. Light and thin, grape seed oil leaves a glossy film over skin when used as a carrier oil for essential oils in aromatherapy. It contains more linoleic acid than many other carrier oils.

Grape seed oil is also used as a lubricant for shaving and as a growth and strengthening treatment for hair.

Potential Medicinal Benefits

Grape seed oil may provide some health benefit. A 1993 study supports the claim that grape seed oil increases high-density lipoprotein (HDL-C or "good cholesterol") levels and reduces LDL levels.

Although grape seeds contain antioxidants and other biologically active compounds, the cold-pressed grape seed oil contains negligible amounts due to their insolubility in lipids. For instance, sufficiently high amounts of resveratrol occur in grape seed for it to be extracted commercially, yet it is almost entirely absent in the grape seed oil.

Potential Medicinal Complications

Grapeseed oil has sometimes been found to contain dangerous levels of polycyclic aromatic hydrocarbons because of direct contact with combustion gases during the drying process.

Composition

Figure: *Grape seeds (numbers 7 and 8) and grapes*

The following table lists a typical fatty acid composition of grape seed oil:

Acid	*Type*	*Percentage*
Linoleic acid	ω–6 unsaturated	69.6%
Oleic acid	ω–9 unsaturated	15.8%
Palmitic acid(Hexadecanoic acid)	Saturated	7%
Stearic acid(Octadecanoic acid)	Saturated	4%
Alpha-linolenic acid	ω–3 unsaturated	0.1%
Palmitoleic acid(9-Hexadecenoic acid)	ω–7 unsaturated	less than 1%

Grape seed oil also contains 0.8 to 1.5% unsaponifiables rich in phenols (tocopherols) and steroids (campesterol, beta-sitosterol, stigmasterol). Grapeseed oil contains small amounts of vitamin E, but safflower oil, cottonseed oil, or rice bran oil contain greater amounts. Grapeseed oil is high in polyunsaturates and low in saturated fat.

Walnut Oil

Walnut oil is oil extracted from English walnuts (also known as Persian walnuts). Each 100.0g of oil provides about 63.3g polyunsaturated fatty acids, 22.8g monounsaturated fats, and 9.1g saturated fats.

It contains no cholesterol. It is about 22.2% monounsaturated oleic acid, an omega-9 fatty acid, 52.9% linoleic acid, an essential polyunsaturated omega-6 fatty acid, and 10.4% alpha linolenic acid, which supplies the body with needed omega-3 fatty acids, another form of polyunsaturated fatty acid essential to human nutrition.

Culinary Use

Figure: *A cup of California roasted walnut oil*

Walnut oil is edible and is generally used less than other oils in food preparation, often due to high pricing. It is light-coloured and delicate in flavour and scent, with a nutty quality.

Although chefs sometimes use walnut oil for pan frying, most avoid walnut oil for high temperature cooking; heating tends to reduce the oil's flavour & nutritive value and to produce a slight bitterness. In addition cooking rapidly destroys the oil's antioxidants. Walnut oil is at its most valuable in cold dishes such as salad dressings, where it lends its flavour to best advantage.

Over 99% of walnut oil sold in the USA is produced in California. There are also producers in France, Australia and New Zealand.

There are two types of walnut oil commercially available: cold pressed, and refined. Cold pressed walnut oil is typically more expensive due to the loss of a higher percentage of the oil. Refined walnut oil is expeller pressed and saturated with solvent to extract the highest percentage of oil available in the nut meat. The solvents are subsequently eliminated by heating the mixture to around 400 degrees F. Both methods produce food grade culinary oils. However, cold pressing preserves more of the nutrients and flavour.

Artistic Use

Walnut oil was one of the most important oils used by Renaissance painters. Its short drying time and lack of yellow tint make it a good oil paint base thinner and brush cleaner.

However, some practitioners consider walnut oil paint film to be inferior to linseed oil paint film. Walnut oil, like all nut seed and vegetable oils will rancidify and rancidification is accelerated by heat, light, and oxygen. Instead of walnut oil, many artists and stores sell linseed oil, poppyseed oil, and safflower oil as replacements.

Some woodworkers favour walnut oil as a finish for implements that will come in contact with food, such as cutting boards and wooden bowls because of their belief in its safety. People who mix oil & wax to formulate wood finishes value walnut oil as an ingredient because of the edibility of both ingredients.

The oil typically is combined with beeswax in a mixture of 1/3 oil to 2/3 beeswax. Others will use mineral oil or its other name butcher block oil. Even though mineral oil is a petroleum based product it is inert and is also used as a laxative.

Rice Bran Oil

Rice bran oil is the oil extracted from the hard outer brown layer of rice after chaff (rice husk). It is notable for its high smoke point of 232 °C (450 °F) and its mild flavour, making it suitable for high-temperature cooking methods such as stir frying and deep frying. It is popular as a cooking oil in several Asian countries, including Japan, India and China.

Uses

Rice bran wax, obtained from rice bran oil, is used as a substitute for carnauba wax in cosmetics, confectionery, shoe creams and polishing compounds. It is an edible oil which is used in the preparation of vegetable ghee.

Composition

Rice bran oil has a composition similar to that of peanut oil, with 38% monounsaturated, 37% polyunsaturated, and 25% saturated fatty acids. The fatty acid composition is:

Fatty acid	*Percentage*
C14:0 Myristic acid	0.6%
C16:0 Palmitic acid	21.5%
C18:0 Stearic acid	2.9%
C18:1 Oleic acid (an Omega 9 fatty acid)	38.4%
C18:2 Linoleic acid (LA, an Omega 6 fatty acid)	34.4%
C18:3 a-Linolenic acid (ALA, an Omega 3 fatty acid)	2.2%

Physical Propreties of Crude & refined Rice bran oil

character	*Crude Rice bran oil*	*Refined oil*
Moisture	0.5-1.0%	0.1-0.15%
Density (15-15 °C)	0.913-0.920	0.913-0.920
Refractive Index	1.4672	1.4672
Iodine value	95-100	95-104
Saponification value	187	187
Unsaponifiable matter	4.5-5.5	1.8-2.5
Free fatty acids	5-15%	0.15-0.2%
oryzanol	2.0	1.5-1.8
Tocopherol	0.15	0.05
Colour(Tintometre)	20Y+2.8R	10Y+1.0R

Health Benefits

A component of rice bran oil is the antioxidant γ-oryzanol, at around 2% of crude oil content. Thought to be a single compound when initially isolated, it is now known to be a mixture of steryl and other triterpenyl esters of ferulic acids. Also significant is the relatively high fractions of tocopherols and tocotrienols, together as vitamin E. Rice bran oil is also rich in other phytosterols.

Cholesterol

Literature review shows rice bran oil and its active constituents improve blood cholesterol by reducing total plasma cholesterol and triglycerides, and increasing the proportion of HDL cholesterol. Results of an animal study indicated a 42% decrease in total cholesterol with a 62% drop in LDL cholesterol, when researchers supplemented test subjects' diets with fractionated vitamin E obtained from rice bran oil.

Menopause

One small-scale study of γ-oryzanol, a mixture of chemicals found in rice bran oil, found that 90% of the women had some form of relief from hot flashes after taking a supplement of the purified concentrate for four to six weeks.

Antioxidant Stability

The oryzanol content of the pan heated rice bran oil samples remains approximately the same even when heated at 180ÚC for 8 hours, while a decrease in oryzanol content was reported in the case of microwave heating at the same conditions.

Flaxseed Oil

Linseed oil, also known as flaxseed oil is a colourless to yellowish oil obtained from the dried, ripened seeds of the flax plant (*Linum usitatissimum*). The oil is obtained by pressing, sometimes followed by solvent extraction. Linseed oil is a *drying oil*, meaning it can polymerize into a solid form.

Due to its polymer-forming properties, linseed oil can be used on its own or blended with combinations of other oils, resins or solvents as an impregnator, drying oil finish or varnish in wood finishing, as a pigment binder in oil paints, as a plasticizer and hardener in putty, and in the manufacture of linoleum.

Linseed oil use has declined over the past several decades with increased availability of synthetic alkyd resins—which function similarly but resist yellowing.

Linseed oil is an edible oil in demand as a nutritional supplement, as a source of α-Linolenic acid, an omega-3 fatty acid). In parts of Europe, it is traditionally eaten with potatoes and quark. It is regarded as a delicacy due to its hearty taste, which enhances the flavour of quark, which is otherwise bland.

Chemical Aspects

Linseed oil is a triglyceride, like other fats. Linseed oil is distinctive for its unusually large amount of α-linolenic acid, which has a distinctive reaction with oxygen in air. Specifically, the fatty acids in a typical linseed oil are of the following types:

- The triply unsaturated α-linolenic acid (51.9-55.2%),
- The saturated acids palmitic acid (about 7%) and stearic acid (3.4-4.6%),
- The monounsaturated oleic acid (18.5-22.6%),
- The doubly unsaturated linoleic acid (14.2-17%).

Having a high content of di- and triunsaturated esters, linseed oil is particularly susceptible to polymerization reactions upon exposure to oxygen in air. This polymerization, which is called *drying,* results in the rigidification of the material. The drying process can be so exothermic as to pose a fire hazard under certain circumstances. To prevent premature drying, linseed oil-based products (oil paints, putty) should be stored in air-tight containers.

Like some other drying oils, linseed oil exhibits fluorescence under UV light after degradation.

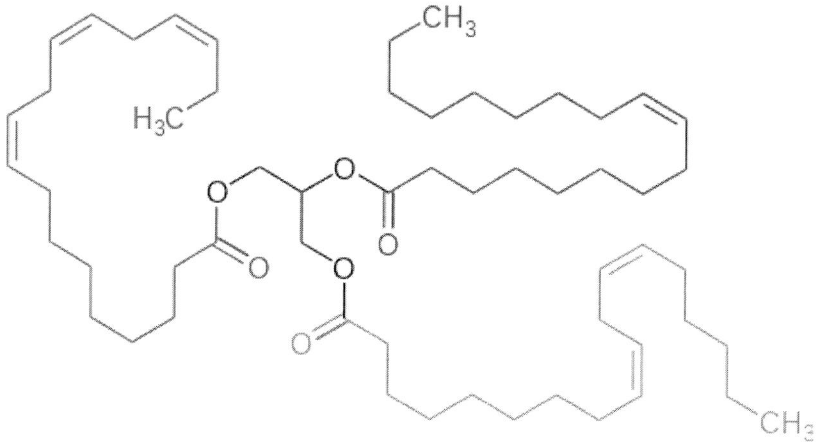

Figure: *Representative triglyceride found in a linseed oil, a triester (triglyceride) derived of linoleic acid, alpha-linolenic acid, and oleic acid.*

Uses

Most applications of linseed oil exploit its drying properties, i.e., the initial material is liquid or at least pliable and the aged material is rigid but not brittle. The water-repelling (hydrophobic) nature of the resulting hydrocarbon-based material is advantageous.

Paint Binder

Linseed oil is a common carrier used in oil paint. It can also be used as a painting medium, making oil paints more fluid, transparent and glossy. It is available in varieties such as cold pressed, alkali refined, sun bleached, sun thickened, and polymerised (stand oil). The introduction of linseed oil was a significant advance in the technology of oil painting.

Putty

Traditional glazing putty, consisting of a paste of chalk powder and linseed oil, is a sealant for glass windows that hardens within a few weeks of application and can then be painted over. The utility of putty is owed to the drying properties of linseed oil.

Wood Finish

When used as a wood finish, linseed oil dries slowly and shrinks little upon hardening. Linseed oil does not cover the surface as varnish does, but soaks into the (visible and microscopic) pores, leaving a shiny but not glossy surface that shows off the grain of the wood. A linseed oil finish is easily repaired, but it provides no significant barrier against scratching.

Only wax finishes are less protective. Liquid water penetrates a linseed oil finish in mere minutes, and water vapour bypasses it almost completely. Garden furniture treated with linseed oil may develop mildew. Oiled wood may be yellowish and is likely to darken with age. Because it fills the pores, linseed oil partially protects wood from denting by compression.

Linseed oil is a traditional finish for gun stocks, though very fine finish may require months to obtain. Several coats of linseed oil is the traditional protective coating for the raw willow wood of cricket bats; it is used so that the wood retains some moisture. New cricket bats are coated with linseed oil and knocked to perfection so they last longer. Linseed oil is also often used by billiards or pool cue-makers for cue shafts, as a lubricant/protectant for wooden recorders, and used in place of epoxy to seal modern wooden surfboards.

Additionally, a luthier may use linseed oil when reconditioning a guitar, mandolin, or other stringed instrument's fret board; lemon-scented mineral oil is commonly used for cleaning, then a light amount of linseed oil (or other drying oil) is applied to protect it from grime that might otherwise result in accelerated deterioration of the wood.

Gilding

Boiled linseed oil is used as sizing in traditional oil gilding to adhere sheets of gold leaf to a substrate (parchment, canvas, Armenian bole, etc.) It has a much longer working time than water-based size and gives a firm smooth surface which is adhesive enough in the first 12–24 hours after application to cause the gold to attach firmly to the intended surface.

Linoleum

Linseed oil is used to bind wood dust, cork particles, and related materials in the manufacture of the floor covering linoleum. After its invention in 1860 by Frederick Walton, linoleum, or 'lino' for short, was a common form of domestic and industrial floor covering from the 1870s until the 1970s when it was largely replaced by PVC ('vinyl') floor coverings. However, since the 1990s, linoleum is on the rise again, being considered more environmentally sound than PVC. Linoleum has given its name to the printmaking technique linocut, in which a relief design is cut into the smooth surface and then inked and used to print an image. The results are similar to those obtained by woodcut printing.

Nutritional Supplement and Food

Flax seed oil is easily oxidized, and rapidly becomes rancid, with an unpleasant odour, unless refrigerated. Even when kept under cool conditions, it has a shelf life of only a few weeks. Oil with an unpleasant or rancid odour should be discarded. Oxidation of flax seed oil is a major commercial concern, and antioxidants may be added to prevent rancidification. Linseed oil is not generally recommended for use in cooking, yet one study does show that the alpha linolenic acid (ALA) while bound in flaxseed was found to be stable for cooking. When bound to flaxseed ALA can withstand temperatures up to 350 degrees F (176.67 C) for two hours.

Food-grade flaxseed oil is cold-pressed, obtained without solvent extraction, in the absence of oxygen, and marketed as edible flaxseed oil. Fresh, refrigerated and unprocessed, linseed oil is used as a nutritional supplement and is a traditional European ethnic food, highly regarded for its hearty taste. It contains the highest level of the omega-3 fatty acid ALA among vegetable oils. Regular flaxseed oil contains between 52% and 63% ALA (C18:3 *n*-3). Plant breeders have developed flaxseed with both higher ALA (70%) and very low ALA content (< 3%). The USFDA granted generally recognized as safe (GRAS) status for high alpha linolenic flaxseed oil.

Nutrient Content

Typical fatty acid content	*%*	*% European*
Palmitic acid	6.0	4.0–6.0
Stearic acid	2.5	2.0–3.0
Arachidic acid	0.5	0–0.5
Palmitoleic acid	-	0–0.5
Oleic acid	19.0	10.0–22.0
Eicosenoic acid	-	0–0.6
Linoleic acid	24.1	12.0–18.0
Alpha-linolenic acid	47.4	56.0–71.0
Other	0.5	-

Nutrition information from the Flax Council of Canada.

Per 1 tbsp (14 g)

- Calories: 126
- Total fat: 14 g
- Omega-3: 8 g
- Omega-6: 2 g
- Omega-9: 3 g

Flax seed oil contains no significant amounts of protein, carbohydrates or fibre.

Additional Uses

- Animal care products
- Bicycle maintenance as a thread fixative, rust inhibitor and lubricant
- Composition ornament for moulded decoration
- Earthen floors
- Animal feeds
- Industrial lubricant
- Leather treatment
- Oilcloth
- Particle Detectors
- Textiles
- Wood preservation (including as an active ingredient of Danish Oil)
- Cookware seasoning

Modified Linseed Oils

Stand Oil

Stand oil is generated by heating linseed oil near 300 °C for a few days in the complete absence of air.

Under these conditions, the polyunsaturated fatty esters convert to conjugated dienes, which then undergo Diels-Alder reactions, leading to crosslinking.

The product, which is highly viscous, gives highly uniform coatings that "dry" to more elastic coatings than linseed oil itself. Soybean oil can be treated similarly, but converts more slowly.

On the other hand, tung oil converts very quickly, being complete in minutes at 260 °C. Coatings prepared from stand oils are less prone to yellowing than are coatings derived from the parent oils.

Boiled Linseed Oil

Today, "boiled linseed oil" refers to a combination of raw linseed oil, stand oil, and metallic dryers (catalysts to accelerate drying). In Medieval times, linseed oil was boiled with lead oxide (litharge) to give a product called boiled linseed oil.

The lead oxide forms lead "soaps" (lead oxide is alkaline) which promotes hardening (polymerisation) of linseed oil by reaction with atmospheric oxygen. Heating shortens its drying time.

Spontaneous Combustion

Rags soaked with linseed oil stored in a pile are considered a fire hazard because they provide a large surface area for oxidation of the oil, and the oil oxidises quickly.

The oxidation of linseed oil is an exothermic reaction, which accelerates as the temperature of the rags increases. When heat accumulation exceeds the rate of heat dissipation into the environment, the temperature increases and may eventually become hot enough to make the rags spontaneously combust.

In 1991, One Meridian Plaza, a high rise in Philadelphia, was severely damaged and three firefighters perished in a fire caused by linseed oil-soaked rags. In 2011, a garage in Sacramento also caught fire due to the spontaneous combustion of linseed oil-soaked rags. In 2014, the Diamond Head home of Hawaii 5-0 actor Alex O'Loughlin was damaged in a similarly started fire.

When the mummy of King Tutankhamen was examined by Howard Carter, he described the mummy as a "charred wreck".

Investigation of the mummy by archeologist Chris Naunton revealed that, with the hasty preparations for the burial, after the body was sealed in its tomb, the linseed oil and linen used in elements of the mummification process resulted in spontaneous combustion that charred the body.

Avocado Oil

Avocado oil is an edible oil pressed from the fruit of the Persea americana (avocado). As a food oil, it is used as an ingredient in other dishes, and as a cooking oil. It is also used for lubrication and in cosmetics, where it is valued for its supposed regenerative and moisturizing properties.

It has an unusually high smoke point, both unrefined and especially when refined. The smoke point of the unrefined form is 480 °F (249 °C) and the refined form can reach 520 °F (271 °C). The exact smoke point depends heavily on the quality of refinement and the way the oil has been handled up until reaching store shelves and subsequent kitchens.

Uses

Avocado oil functions well as a carrier oil for other flavours. It is high in monounsaturated fats and vitamin E. Avocado oil also enhances the absorption of carotenoids and other nutrients.

Because the avocado is a year-round crop, some olive oil processing facilities, particularly in Australia and New Zealand, process olive oil during the olive season, and avocado oil during the rest of the year.

Avocado oil was originally, and still is, extracted for cosmetic use because of its very high skin penetration and rapid absorption. Following drying of the avocado flesh to remove as much water as possible (H"65% water in avocado flesh), avocado oil for cosmetics is traditionally extracted with solvents at elevated temperatures.

After extraction, the oil for application in skin care products is usually refined, bleached, and deodorized, resulting in an odorless yellow oil. Like extra virgin olive oil, cold-pressed avocado oil is unrefined and so retains the flavour and colour characteristics of the fruit flesh.

Properties

Extra virgin avocado oil from the Hass cultivar has a characteristic flavour, is high in monounsaturated fatty acids, and has a high smoke point (e"250 °C or 482 °F), making it a good oil for frying. 'Hass' cold-pressed avocado oil is a brilliant emerald green when extracted; the colour is attributed to high levels of chlorophylls and carotenoids extracted into the oil. Cold-pressed 'Hass' avocado oil has been described as having an avocado flavour, with grassy and butter/mushroom-like flavours. Other varieties may produce oils of slightly different flavour profile as has been seen with 'Fuerte,' which has been described as having more mushroom and less avocado flavour. As a culinary oil, avocado oil compares well with olive oil. It has a similar monounsaturated fat profile which helps to protect the oil from breakdown during heating. Avocado oil is naturally low acidic, helping to increase smoke point.

An extra virgin avocado oil, characterized by a deep emerald green colour (from avocado's chlorophyll content) can safely be heated to degree of 480 °F (249 °C)—similar to a high quality extra virgin olive oil.

Both unrefined and refined avocado oil can safely be used to conduct almost any high heat cooking application including baking, stir-fry, deep-fry, sear, barbecue, roast and saute.

Avocado oil is relatively new to the culinary world and is often mislabelled in regard to smoke point. It is important to note that like all oils, the more refined, the higher the smoke point.

Avocado oil is one of few edible oils not derived from seeds; it is pressed from the fleshy pulp surrounding the avocado pit.

Chapter 7

Food Safety and Quality Issues of Dairy Fats

Milk, according to the PFA Rules, is the normal mammary secretion derived from the complete milking of healthy milch animal without either addition thereto or extraction therefrom. It shall be free from colostrum.

Milk of different classes and of different designations shall conform to the standards laid down.

It shall conform to both the parametres for milk fat and milk solids-not-fat (SNF), independently, as prescribed in columns four and five of this table. Milk shall include cow's or buffalo's, or sheep's or goat's or a mixture thereof.

Table: *PFA standards for different classes and designations of milk in India*

				Minimum percentage
Class of Milk	*Designations*	*State and Union Territories*	*Milk fat*	*Milk solids -not-fat (SNF)*
Buffalo milk	Raw, pasteurized, boiled, flavoured and sterilized	Assam; Bihar; Chandigarh; Delhi; Gujarat; Maharashtra; Haryana;Meghalaya; Punjab; Sikkim; Uttar Pradesh; West Bengal; Andaman & Nicobar; Andhra Pradesh; Arunachal Pradesh; Dadra & Nagar Haveli; Goa; Daman & Diu; Kerala; Himachal Pradesh; Jammu & Kashmir; Karnataka.	6.0	9.0
	-do-	Kerala; Lakshadweep; Madhya Pradesh; Manipur; Mizoram; Nagaland; Orissa; Pondicherry; Rajasthan; Tripura; Tamil Nadu.	5.0	9.0

Contd...

Cow milk	-do-	Chandigarh; Haryana; Punjab.	4.0	8.5
	-do-	Andaman & Nicobar; Andhra Pradesh; Arunachal Pradesh; Assam; Bihar; Dadra and Nagar Haveli; Delhi; Goa; Daman & Diu; Gujarat; Himachal Pradesh; Jammu & Kashmir; Karnataka; Kerala; Lakshadweep; Madhya Pradesh; Maharashtra; Manipur; Meghalaya; Nagaland, Pondicherry; Rajasthan; Sikkim; Tamil Nadu; Tripura; Uttar Pradesh; West Bengal.	3.5	8.5
	-do-	Mizoram; Orissa.	3.0	8.5
Goat or sheep milk	-do-	Chandigarh; Haryana; Kerala; Madhya Pradesh; Maharashtra; Punjab; Uttar Pradesh.	3.5	9.0
	-do-	Andaman & Nicobar; Andhra Pradesh; Arunachal Pradesh; Assam; Bihar; Dadra & Nagar Haveli; Delhi; Goa; Daman & Diu; Gujarat; Himachal Pradesh; Jammu & Kashmir; Karnataka; Lakshadweep; Manipur; Meghalaya; Mizoram; Nagaland; Orissa; Pondicherry; Rajasthan; Sikkim; Tamil Nadu; Tripura; West Bengal.	3.0	9.0
Mixed milk	-do-	All India	4.5	8.5
Standardized milk	Pasteurized, flavoured and sterilized	All India	4.5	8.5
Recombined milk	-do-	All India	3.0	8.5
Toned milk	-do-	All India	3.0	8.5
Double toned milk	-do-	All India	1.5	9.0
Skimmed milk	-do-	All India	Not more than 0.5	8.7
Full cream milk	Pasteurized and sterilized	All India	6.0	9.0

The *Codex Alimentarius* (Latin for "Book of Food") is a collection of internationally recognized standards, codes of practice, guidelines and other recommendations relating to foods, food production and food safety.

Its name is derived from the Codex Alimentarius Austriacus. Its texts are developed and maintained by the *Codex Alimentarius* Commission, a body that was established in early November 1961 by the Food and Agriculture Organization of the United Nations (FAO), was joined by the World Health Organization (WHO) in June 1962, and held its first session in Rome in October 1963. The Commission's main goals are to protect the health of consumers and ensure fair practices in the international food trade. The Codex Alimentarius is recognized by the World Trade Organization as an international reference point for the resolution of disputes concerning food safety and consumer protection.

As of 2012, there were the 186 members of the Codex Alimentarius Commission: 186 member countries and one member organization, theEuropean Union (EU). There were 215 Codex observers: 49 intergovernmental organizations, 150 non-governmental organizations, and 16United Nations organizations.

Scope

The Codex Alimentarius covers all foods, whether processed, semi-processed or raw. In addition to standards for specific foods, the Codex Alimentarius contains general standards covering matters such as food labelling, foodhygiene, food additives and pesticide residues, and procedures for assessing the safety of foods derived from modern biotechnology. It also contains guidelines for the management of official i.e. governmental import and exportinspection and certification systems for foods.

The Codex Alimentarius is published in Arabic, Chinese, English, French, Spanish and Russian. Not all texts are available in all languages.

General Texts

- Food labelling (general standard, guidelines on nutrition labelling, guidelines on labelling claims)
- Food additives (general standard including authorized uses, specifications for food grade chemicals)
- Contaminants in foods (general standard, tolerances for specific contaminants including radionuclides, aflatoxins and other mycotoxins)
- Pesticide and veterinary chemical residues in foods (maximum residue limits)
- Risk assessment procedures for determining the safety of foods derived from biotechnology (DNA-modified plants, DNA-modified micro-organisms, allergens)
- Food hygiene (general principles, codes of hygienic practice in specific industries or food handling establishments, guidelines for the use of the Hazard Analysis and Critical Control Point or "HACCP" system)
- Methods of analysis and sampling

Specific Standards

- Meat products (fresh, frozen, processed meats and poultry)
- Fish and fishery products (marine, fresh water and aquaculture)
- Milk and milk products
- Foods for special dietary uses (including infant formula and baby foods)
- Fresh and processed vegetables, fruits, and fruit juices
- Cereals and derived products, dried legumes
- Fats, oils and derived products such as margarine
- Miscellaneous food products (chocolate, sugar, honey, mineral water)

The European Food Safety Authority (EFSA)

The European Food Safety Authority (EFSA) is the agency of the European Union (EU) that provides independent scientific advice and communicates on existing and emerging risks associated with the food chain. EFSA was established in February 2002, is based in Parma, Italy and had a budget for 2008 of € 65.9 million.

The work of EFSA covers all matters with a direct or indirect impact on food and feed safety, including animal health and welfare, plant protection and plant health and nutrition. EFSA supports the European Commission, the European Parliament and EU member states in taking effective and timely risk management decisions that ensure the protection of the health of European consumers and the safety of the food and feed chain. EFSA also communicates to the public in an open and transparent way on all matters within its remit.

The Importance of the HACCP in Food Production

Hazard analysis and critical control points or HACCP is a systematic preventive approach to food safety from biological, chemical, and physical hazards in production processes that can cause the finished product to be unsafe, and designs measurements to reduce these risks to a safe level.

In this manner, HACCP is referred as the prevention of hazards rather than finished product inspection. The HACCP system can be used at all stages of a food chain, from food production and preparation processes including packaging, distribution, etc. The Food and Drug Administration (FDA) and theUnited States Department of Agriculture (USDA) say that their mandatory HACCP programmes for juice and meat are an effective approach to food safety and protecting public health. Meat HACCP systems are regulated by the USDA, while seafood and juice are regulated by the FDA. The use of HACCP is currently voluntary in other food industries.

HACCP is believed to stem from a production process monitoring used duringWorld War II because traditional "end of the pipe" testing on artillery shell's firing mechanisms could not be performed, and a large percentage of the artillery shells made at the time were either duds or misfiring. HACCP itself was conceived in the 1960s when the US National Aeronautics and Space Administration (NASA) asked Pillsbury to design and manufacture the first foods for space flights. Since then, HACCP has been recognized internationally as a logical tool for adapting traditional inspection methods to a modern, science-based, food safety system. Based on risk-assessment, HACCP plans allow both industry and government to allocate their resources efficiently in establishing and auditing safe food production practices. In 1994, the organization of *International HACCP Alliance* was established initially for the US meat and poultry industries to assist them with implementing HACCP and now its membership has been spread over other professional/industrial areas.

Hence, HACCP has been increasingly applied to industries other than food, such as cosmetics and pharmaceuticals. This method, which in effect seeks to plan out unsafe practices based on science, differs from traditional "produce and sort" quality control methods that do nothing to prevent hazards from occurring and must identify them at the end of the process. HACCP is focused only on the health safety issues of a product and not the quality of the product, yet HACCP principles are the basis of most food quality and safety assurance systems, and the United States, HACCP compliance is regulated by 21 CFRpart 120 and 123. Similarly, FAO/WHO published a guideline for all governments to handle the issue in small and less developed food businesses.

Principles

1. Conduct a hazard analysis: Plans determine the food safety hazards and identify the preventive measures the plan can apply to control these hazards. A food safety hazard is any biological, chemical, or physical property that may cause a food to be unsafe for human consumption.
2. Identify critical control points: A critical control point (CCP) is a point, step, or procedure in a food manufacturing process

at which control can be applied and, as a result, a food safety hazard can be prevented, eliminated, or reduced to an acceptable level.

3. Establish critical limits for each critical control point: A critical limit is the maximum or minimum value to which a physical, biological, or chemical hazard must be controlled at a critical control point to prevent, eliminate, or reduce to an acceptable level.
4. Establish critical control point monitoring requirements: Monitoring activities are necessary to ensure that the process is under control at each critical control point. In the United States, the FSIS requires that each monitoring procedure and its frequency be listed in the HACCP plan.
5. Establish corrective actions: These are actions to be taken when monitoring indicates a deviation from an established critical limit. The final rule requires a plant's HACCP plan to identify the corrective actions to be taken if a critical limit is not met. Corrective actions are intended to ensure that no product is injurious to health or otherwise adulterated as a result of the deviation enters commerce.
6. Establish procedures for ensuring the HACCP system is working as intended: Validation ensures that the plants do what they were designed to do; that is, they are successful in ensuring the production of a safe product. Plants will be required to validate their own HACCP plans. FSIS will not approve HACCP plans in advance, but will review them for conformance with the final rule.

Verification ensures the HACCP plan is adequate, that is, working as intended. Verification procedures may include such activities as review of HACCP plans, CCP records, critical limits and microbial sampling and analysis. FSIS is requiring that the HACCP plan include verification tasks to be performed by plant personnel. Verification tasks would also be performed by FSIS inspectors. Both FSIS and industry will undertake microbial testing as one of several verification activities.

Verification also includes 'validation' – the process of finding evidence for the accuracy of the HACCP system (e.g. scientific evidence for critical limitations).

7. Establish record keeping procedures: The HACCP regulation requires that all plants maintain certain documents, including

its hazard analysis and written HACCP plan, and records documenting the monitoring of critical control points, critical limits, verification activities, and the handling of processing deviations. Implementation involves monitoring, verifying, and validating of the daily work that is compliant with regulatory requirements in all stages all the time. The differences among those three types of work are given by Saskatchewan Agriculture and Food.

Standards

The seven HACCP principles are included in the international standard ISO 22000 FSMS 2005. This standard is a complete food safety and quality management system incorporating the elements of prerequisite programmes(GMP & SSOP), HACCP and the quality management system, which together form an organization's Total Quality Management system.

Other Standards, such as Safe Quality Food Institute's SQF 2000 Code, also relies upon the HACCP methodology as the basis for developing and maintaining Food Safety (Level 2) and Food Quality (Level 3) Plans and programmes in concert with the fundamental prerequisites of Good Manufacturing Practices.

Training

Training for developing and implementing HACCP Food Safety management system are offered by several quality assurance companies. However, ASQ does provide a Trained HACCP Auditor (CHA) exam to individuals seeking professional training. In the UK the Chartered Institute of Environmental Health (CIEH) offers a HACCP for Food Manufacturing qualification, accredited by the QCA (Qualifications and Curriculum Authority).

Application

- Fish and fishery products
- Fresh-cut produce
- Drinks and nectary products
- Food outlets
- Meat and poultry products
- School food and services

Water Quality Management

The use of HACCP for water quality management was first proposed nearly 20 years ago. Thereafter, a number of water quality initiatives applied HACCP principles and steps to the control of infectious disease from water, and provided the basis for the Water Safety Plan (WSP) approach in the third edition of the World Health Organization (WHO) report, which has been described as "a way of adapting the HACCP approach to drinking water systems".

HACCP for Building Water Systems

Hazards associated with water systems in buildings include physical, chemical and microbial hazards. In 2013, NSF International, a public health and safety NGO, established education, training and certification programmes in HACCP for Building Water Systems. The programmes, developed with the guidance of subject matter experts Aaron Rosenblatt and William McCoy centre on the use of the use of HACCP principles adapted to the specific requirements of domestic (hot and cold) and utility (HVAC, etc.) water systems in buildings, in order to prevent plumbing-associated hazards from harming people. Hazards addressed include scalding, lead, and disinfection byproductsas well as a range of clinically important pathogens, such as Legionella, pseudomonas, non-tubercular Mycobacteria (NTM), acinetobacter, Elizabethkingia and Naegleria. Early adopters of HACCP for Building Water Systems include leading healthcare institutions, notably Mayo Clinic (Rochester, MN).

ISO 22000

ISO 22000 is like a standard designed to help augment HACCP on issues related to food safety. Although several companies, especially the big ones, have either implemented or are on the point of implementing ISO 22000, there are many others which are hesitant to adopt it. The main reason behind that is the lack of information and the fear that the new standard is too demanding in terms of bureaucratic work, from abstract of case study.

Food Safety Standards

The Food Safety and Standards Authority of India (FSSAI) is an agency of the Ministry of Health & Family Welfare, Government of India. The FSSAI is responsible for protecting and promoting public health through the regulationand supervision of food safety. The FSSAI has been established under the Food Safety and Standards Act, 2006 which is a consolidating statute related to food safety and regulation in India.

The FSSAI is led by a non-executive Chairperson, appointed by the Central Government from amongst the persons of eminence in the field of food science or from amongst the persons from the administration who have been associated with the subject and is either holding or has held the position of not below the rank of Secretary to the Government of India. Mr. K.Chandramouli is the current Chairperson.

The executive head of FSSAI is the Chief Executive Officer .The FSSAI has its headquarters at New Delhi. The agency also has 5 regional offices located in Delhi, Guwahati, Mumbai, Kolkata and Chennai, 4 referral laboratories and 72 local laboratories located throughout India.

The International Food Standard (IFS) is a common food safety standard with a uniform evaluation system used to qualify and select suppliers. It helps retailers ensure the food safety of their products and monitors the quality level of producers of retailer branded food products.

Having a common standard and a uniform way to evaluate the quality level of suppliers reduces a retailer's need for second party audits. For the retailer, supply chain management becomes more consistent and efficient. It reduces the overall costs of the process, and increases the level of safety for customers, suppliers, and consumers. While it may be a ticket to trade with retailers, it is also a chance for you to demonstrate your commitment to food safety, quality, and legality, and to work on continual improvement.

What is the Standard?

Concern among retailers about food safety because of their direct responsibility in case of an incident led to the development of uniform food safety standards and audit system. IFS (International Food Standard) was developed by German and French food trade associations to take control of the situation, and it is the international standard for auditing retailer and wholesaler branded food products. It is also supported by the major Italian trade association.

The standard applies to suppliers at all steps of the food processing subsequent to the agricultural stage. The requirements are related to the quality management system and the HACCP system, supported by detailed prerequisite programmes, that is a set of GMP (Good Manufacturing Practice), GLP (Good Laboratory Practice) and GHP (Good Hygiene Practice) requirements. It helps ensure that they comply with quality and safety demands, and legal requirements.

IFS meets the criteria of the Global Food Safety Initiative by CIES – The Food Business Forum, the global organisation including

CEOs and senior management of around 400 retailers (operating close to 200,000 stores) and manufacturer members of all sizes.

The most important European retailers require IFS certification from the suppliers that are included in their supply chain. But IFS is an international standard and can be applied to suppliers outside Europe as well.

The standard enables organisation to:

- Provide evidence of commitment, and, in case of a food safety incident, legal defence in the frame of the "due diligence" concept*
- Build and operate a management system capable of helping you to better meet food quality/safety requirements and legal compliance, with specific reference to the legislation applicable in the countries where the finished product is consumed
- Provide a tool for food safety performance improvement, and the means to monitor and measure food safety performance effectively
- Facilitate reductions in product waste, product reworking, and product recall

Companies that plan well will often have an advantage and be better prepared for certification. It is important that you and your company are positive, committed, and set clear target dates for implementation and assessment.

This includes:

- Establishing a quality management system
- Identifying legal requirements
- Identifying and documenting the specific food safety hazards and the relevant control measures (HACCP system)
- Identifying the applicable Good Manufacturing Practice/Good Hygiene Practice, including a pest control programme, equipment and building maintenance programme, housekeeping and cleaning programme, and all the specific standard requirements
- Implement any needed structural improvements

Food-borne Disease: the Problem

Foodborne illness (also foodborne disease and colloquially referred to asfood poisoning) is any illness resulting from the consumption ofcontaminated food, pathogenic bacteria, viruses, or parasites that contaminate food, as well as chemical or natural toxins such as poisonous mushrooms.

Symptoms vary depending on the cause, and are described below in this article. A few broad generalizations can be made, e.g.: The incubation periodranges from hours to days, depending on the cause and on how much was consumed. The incubation period tends to cause sufferers to not associate the symptoms with the item consumed, and so to cause sufferers to attribute the symptoms to gastroenteritis for example. Symptoms often include vomiting, fever, and aches, and may include diarrhea. Bouts of vomiting can be repeated with an extended delay in between, because even if infected food was eliminated from the stomach in the first bout, microbes (if applicable) can pass through the stomach into the intestine via cells lining the intestinal walls and begin to multiply. Some types of microbes stay in the intestine, some produce a toxin that is absorbed into the bloodstream, and some can directly invade deeper body tissues.

Causes

Figure: *Poorly stored food in a refrigerator*

Foodborne illness usually arises from improper handling, preparation, or food storage. Good hygiene practices before, during, and after food preparation can reduce the chances of contracting an illness. There is a consensus in the public health community that regular hand-washing is one of the most effective defences against the spread of foodborne illness. The action of monitoring food to ensure that it will not cause foodborne illness is known as food safety. Foodborne disease can also be caused by a large variety of toxins that affect the environment. Foodborne illness can also be caused by pesticides or medicines in food and naturally toxic substances such as poisonous mushrooms or reef fish.

Bacteria

Bacteria are a common cause of foodborne illness. In the United Kingdom during 2000, the individual bacteria involved were the following: *Campylobacter jejuni*77.3%, *Salmonella* 20.9%, *Escherichia coli O157:H7* 1.4%, and all others less than 0.56%. In the past, bacterial infections were thought to be more prevalent because few places had the capability to test for norovirus and no active surveillance was being done for this particular agent.

Toxins from bacterial infections are delayed because the bacteria need time to multiply. As a result symptoms associated with intoxication are usually not seen until 12–72 hours or more after eating contaminated food. Usually the symptoms are seen the day after the food has been ingested and digested completely. However if the intoxication involves preformed toxins as is the case with Staphylococcal food poisoning, the symptoms appear within a few hours.

Most common bacterial foodborne pathogens are:

- *Campylobacter jejuni* which can lead to secondary Guillain–Barré syndrome and periodontitis
- *Clostridium perfringens*, the "cafeteria germ"
- *Salmonella* spp. – its *S. typhimurium* infection is caused by consumption of eggs or poultry that are not adequately cooked or by other interactive human-animal pathogens

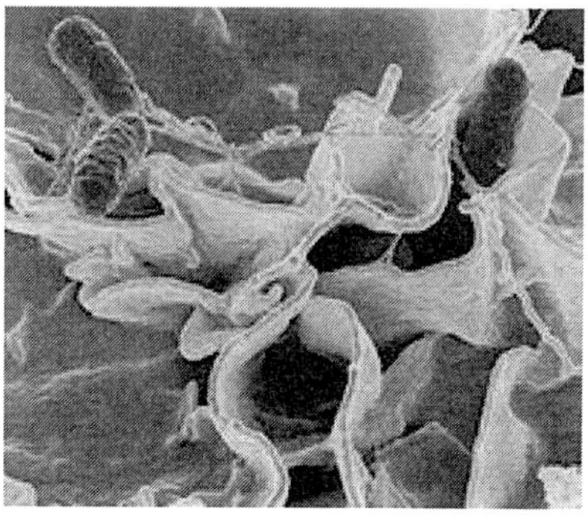

Figure: *Salmonella*

- *Escherichia coli O157:H7* enterohemorrhagic (EHEC) which can cause hemolytic-uremic syndrome

Other common bacterial foodborne pathogens are:

- *Bacillus cereus*
- *Escherichia coli*, other virulence properties, such as enteroinvasive (EIEC), enteropathogenic (EPEC), enterotoxigenic (ETEC), enteroaggregative (EAEC or EAgEC)
- *Listeria monocytogenes*
- *Shigella* spp.
- *Staphylococcus aureus*
- *Staphylococcal enteritis*
- *Streptococcus*
- *Vibrio cholerae*, including O1 and non-O1
- *Vibrio parahaemolyticus*
- *Vibrio vulnificus*
- *Yersinia enterocolitica* and *Yersinia pseudotuberculosis*

Less common bacterial agents:

- *Brucella* spp.
- *Corynebacterium ulcerans*
- *Coxiella burnetii* or Q fever
- *Plesiomonas shigelloides*

Enterotoxins

In addition to disease caused by direct bacterial infection, some foodborne illnesses are caused by enterotoxins (exotoxinstargeting the intestines). Enterotoxins can produce illness even when the microbes that produced them have been killed.

Symptom appearance varies with the toxin but may be rapid in onset, as in the case of enterotoxins of *Staphylococcus aureus* in which symptoms appear in one to six hours. This causes intense vomiting including or not including diarrhea (resulting in staphylococcal enteritis), and staphylococcal enterotoxins (most commonly staphylococcal enterotoxin A but also including staphylococcal enterotoxin B) are the most commonly reported enterotoxins although cases of poisoning are likely underestimated.

It occurs mainly in cooked and processed foods due to competition with other biota in raw foods, and humans are the main cause of contamination as a substantial percentage of humans are persistent carriers of *S. aureus*. The CDC has estimated about 240,000 cases per year in the United States.

- *Clostridium botulinum*
- *Clostridium perfringens*
- *Bacillus cereus*

The rare but potentially deadly disease botulism occurs when the anaerobic bacterium *Clostridium botulinum* grows in improperly canned low-acid foods and produces botulin, a powerful paralytic toxin.

Pseudoalteromonas tetraodonis, certain species of Pseudomonas and Vibrio, and some other bacteria, produce the lethaltetrodotoxin, which is present in the tissues of some living animal species rather than being a product of decomposition.

Emerging Foodborne Pathogens

Many foodborne illnesses remain poorly understood. Approximately 99.9% percent of outbreaks are caused by unknown sources.

- *Aeromonas hydrophila, Aeromonas caviae, Aeromonas sobria*

Preventing Bacterial Food Poisoning

Prevention is mainly the role of the state, through the definition of strict rules ofhygiene and a public services of veterinary surveying of animal products in the food chain, from farming to the transformation industry and delivery (shops andrestaurants). This regulation includes:

- traceability: in a final product, it must be possible to know the origin of the ingredients (originating farm, identification of the harvesting or of the animal) and where and when it was processed; the origin of the illness can thus be tracked and solved (and possibly penalized), and the final products can be removed from the sale if a problem is detected;
- enforcement of hygiene procedures such as HACCP and the "cold chain";
- power of control and of law enforcement of veterinarians.

In August 2006, the United States Food and Drug Administration approved Phage therapy which involves spraying meat with viruses that infect bacteria, and thus preventing infection. This has raised concerns, because without mandatory labelling consumers would not be aware that meat and poultry products have been treated with the spray.

At home, prevention mainly consists of good food safety practices. Many forms of bacterial poisoning can be prevented even if food is contaminated by cooking it sufficiently, and either eating it quickly or refrigerating it effectively. Many toxins, however, are not destroyed by heat treatment.

Techniques that help prevent food borne illness in the kitchen are hand washing, preventing cross-contamination, proper storage, and maintaining cooking temperatures. In general, freezing or refrigerating prevents virtually all bacteria from growing, and heating food sufficiently kills parasites, viruses, and most bacteria. The "danger zone" is a zone of temperatures in which bacteria grow most rapidly, which is 40 °F-140 °F. Storing food below or above the "danger zone" can effectively limit the production of toxins. For storing leftovers, the food must be put in shallow containers for quick cooling and must be refrigerated within two hours. When food is reheated, it must reach an internal temperature of 165 °C (329 °F) or until hot or steaming to kill bacteria.

Mycotoxins and Alimentary Mycotoxicoses

The term alimentary mycotoxicoses refers to the effect of poisoning by Mycotoxins (The term 'mycotoxin' is usually reserved for the toxic chemical products produced by fungi that readily colonize crops) through food consumption.Mycotoxins sometimes have important effects on human and animal health. For example, an outbreak which occurred in the UK in 1960 caused the death of 100,000 turkeys which had consumed aflatoxin-contaminated peanut meal. In the USSR in World War II, 5,000 people died due to Alimentary Toxic Aleukia (ALA). The common foodborne Mycotoxinsinclude:

- Aflatoxins – originated from Aspergillus parasiticus and Aspergillus flavus. They are frequently found in tree nuts, peanuts, maize, sorghum and other oilseeds, including corn and cottonseeds. The pronounced forms of Aflatoxins are those of B1, B2, G1, and G2, amongst which Aflatoxin B1 predominantly targets the liver, which will result in necrosis, cirrhosis, and carcinoma. In the US, the acceptable level of total aflatoxins in foods is less than 20 μg/kg, except for Aflatoxin M1 in milk, which should be less than 0.5 μg/kg. The official document can be found at FDA's website.
- Altertoxins – are those of Alternariol (AOH), Alternariol methyl ether (AME), Altenuene (ALT), Altertoxin-1 (ATX-1), Tenuazonic acid (TeA) and Radicinin (RAD), originated

from Alternaria spp. Some of the toxins can be present in sorghum, ragi, wheat and tomatoes. Some research has shown that the toxins can be easily cross-contaminated between grain commodities, suggesting that manufacturing and storage of grain commodities is a critical practice.

- Citrinin
- Citreoviridin
- Cyclopiazonic acid
- Cytochalasins
- Ergot alkaloids / Ergopeptine alkaloids – Ergotamine
- Fumonisins – Crop corn can be easily contaminated by the fungi Fusarium moniliforme, and its Fumonisin B1 will cause Leukoencephalomalacia (LEM) in horses, Pulmonary edema syndrome (PES) in pigs, liver cancer in rats andEsophageal cancer in humans. For human and animal health, both the FDA and the EC have regulated the content levels of toxins in food and animal feed.
- Fusaric acid
- Fusarochromanone
- Kojic acid
- Lolitrem alkaloids
- Moniliformin
- 3-Nitropropionic acid
- Nivalenol
- Ochratoxins – In Australia, The Limit of Reporting (LOR) level for Ochratoxin A (OTA) analyses in 20th Australian Total Diet Survey was 1 μg/kg, whereas the EC restricts the content of OTA to 5 μg/kg in cereal commodities, 3 μg/kg in processed products and 10 μg/kg in dried vine fruits.
- Oosporeine
- Patulin – Currently, this toxin has been advisably regulated on fruit products. The EC and the FDA have limited it to under 50 μg/kg for fruit juice and fruit nectar, while limits of 25 μg/kg for solid-contained fruit products and 10 μg/kg for baby foods were specified by the EC.
- Phomopsins
- Sporidesmin A
- Sterigmatocystin

- Tremorgenic mycotoxins – Five of them have been reported to be associated with molds found in fermented meats. These are Fumitremorgen B, Paxilline, Penitrem A, Verrucosidin, and Verruculogen.
- Trichothecenes – sourced from Cephalosporium, Fusarium, Myrothecium, Stachybotrys and Trichoderma. The toxins are usually found in molded maize, wheat, corn, peanuts and rice, or animal feed of hay and straw. Four trichothecenes, T-2 toxin, HT-2 toxin, diacetoxyscirpenol (DAS) and deoxynivalenol (DON) have been most commonly encountered by humans and animals. The consequences of oral intake of, or dermal exposure to, the toxins will result in Alimentary toxic aleukia, neutropenia, aplastic anemia, thrombocytopenia and/or skin irritation. In 1993, theFDA issued a document for the content limits of DON in food and animal feed at an advisory level. In 2003, US published a patent that is very promising for farmers to produce a trichothecene-resistant crop.
- Zearalenone
- Zearalenols

Viruses

cases of food poisoning in developed countries. In the US, more than 50% of cases are viral and noroviruses are the most common foodborne illness, causing 57% of outbreaks in 2004. Foodborne viral infection are usually of intermediate (1–3 days) incubation period, causing illnesses which are self-limited in otherwise healthy individuals; they are similar to the bacterial forms described above.

- Enterovirus
- Hepatitis A is distinguished from other viral causes by its prolonged (2–6 week) incubation period and its ability to spread beyond the stomach and intestines into the liver. It often results in jaundice, or yellowing of the skin, but rarely leads to chronic liver dysfunction. The virus has been found to cause infection due to the consumption of fresh-cut produce which has fecal contamination.
- Hepatitis E
- Norovirus
- Rotavirus

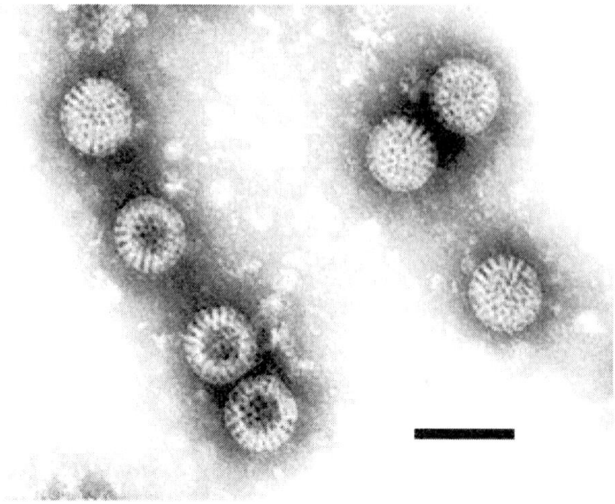

Figure: *Rotavirus*

Parasites

Most foodborne parasites are zoonoses.

- Platyhelminthes:
- *Diphyllobothrium* sp.
- *Nanophyetus* sp.
- *Taenia saginata*
- *Taenia solium*
- *Fasciola hepatica*
- Nematode:
- *Anisakis* sp.
- *Ascaris lumbricoides*
- *Eustrongylides* sp.
- *Trichinella spiralis*
- *Trichuris trichiura*
- Protozoa:
- *Acanthamoeba* and other free-living amoebae
- *Cryptosporidium parvum*
- *Cyclospora cayetanensis*
- *Entamoeba histolytica*
- *Giardia lamblia*

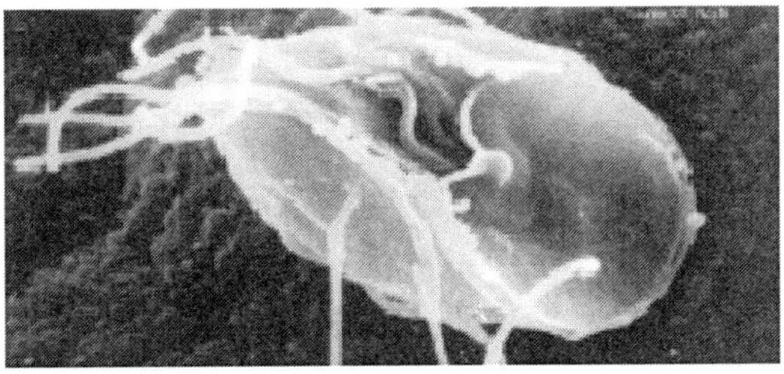

Figure: *Giardia lamblia*

- *Sarcocystis hominis*
- *Sarcocystis suihominis*
- *Toxoplasma gondii*

Natural Toxins

Several foods can naturally contain toxins, many of which are not produced by bacteria. Plants in particular may be toxic; animals which are naturally poisonous to eat are rare. In evolutionary terms, animals can escape being eaten by fleeing; plants can use only passive defences such as poisons and distasteful substances, for example capsaicin in chili peppers and pungent sulfur compounds in garlic and onions. Most animal poisons are not synthesised by the animal, but acquired by eating poisonous plants to which the animal is immune, or by bacterial action.

- Alkaloids
- Ciguatera poisoning
- Grayanotoxin (honey intoxication)
- Mushroom toxins
- Phytohaemagglutinin (red kidney bean poisoning; destroyed by boiling)
- Pyrrolizidine alkaloids
- Shellfish toxin, including paralytic shellfish poisoning, diarrhetic shellfish poisoning, neurotoxic shellfish poisoning,amnesic shellfish poisoning and ciguatera fish poisoning
- Scombrotoxin
- Tetrodotoxin (fugu fish poisoning)

Some plants contain substances which are toxic in large doses, but have therapeutic properties in appropriate dosages.

- Foxglove contains cardiac glycosides.
- Poisonous hemlock (conium) has medicinal uses.

Other Pathogenic Agents

- Prions, resulting in Creutzfeldt–Jakob disease

"Ptomaine Poisoning"

An early theory on the causes of food poisoning involved *ptomaines* (from Greek *ptôma*, "fall, fallen body, corpse"),alkaloids found in decaying animal and vegetable matter, especially (as reflected in their names) putrescine andcadaverine.

While some alkaloids do cause poisoning, the discovery of bacteria left the ptomaine theory obsolete, though as recently as 1882 the Merck's Bulletin stated, "We name such products of bacterial origin ptomaines; and the special alkaloid produced by the comma bacillus is variously named Cadaverine, Putrescine, etc."

Tainted potato salad sickening hundreds at a Communist political convention in Massillon, Ohio, and aboard a Washington, DC cruise boat in separate incidents during a single week drew national attention to the dangers of so-called "ptomaine poisoning" in the pages of the American news weekly, *Time.* Another newspaper article from 1944 told of more than 150 persons being hospitalized in Chicago with ptomaine poisoning apparently from rice pudding served by a chain of restaurants.

Mechanism

Incubation Period

The delay between consumption of a contaminated food and appearance of the first symptoms of illness is called theincubation period. This ranges from hours to days (and rarely months or even years, such as in the case of listeriosis or Creutzfeldt–Jakob disease), depending on the agent, and on how much was consumed. If symptoms occur within one to six hours after eating the food, it suggests that it is caused by a bacterial toxin or a chemical rather than live bacteria.

The long incubation period of many foodborne illnesses tends to cause sufferers to attribute their symptoms to stomach flu.

During the incubation period, microbes pass through the stomach into the intestine, attach to the cells lining the intestinal walls, and begin to multiply there. Some types of microbes stay in the intestine, some produce a toxin that is absorbed into the bloodstream, and some can directly invade the deeper body tissues. The symptoms produced depend on the type of microbe.

Infectious Dose

The infectious dose is the amount of agent that must be consumed to give rise to symptoms of foodborne illness, and varies according to the agent and the consumer's age and overall health. In the case of *Salmonella* a relatively large inoculum of 1 million to 1 billion organisms is necessary to produce symptoms in healthy human volunteers, as*Salmonellae* are very sensitive to acid. An unusually high stomach pH level (low acidity) greatly reduces the number of bacteria required to cause symptoms by a factor of between 10 and 100.

Organizations

The World Health Organization Department of Food Safety and Zoonoses (FOS) provides scientific advice for organizations and the public on issues concerning the safety of food. Its mission is to lower the burden of foodborne disease, thereby strengthening the health security and sustainable development of Member States. Foodborne and waterborne diarrhoeal diseases kill an estimated 2.2 million people annually, most of whom are children. WHO works closely with the Food and Agriculture Organization of the United Nations (FAO) to address food safety issues along the entire food production chain—from production to consumption—using new methods of risk analysis. These methods provide efficient, science-based tools to improve food safety, thereby benefiting both public health and economic development.

International Food Safety Authorities Network (INFOSAN)

The International Food Safety Authorities Network (INFOSAN) is a joint programme of the WHO and FAO. INFOSAN has been connecting national authorities from around the globe since 2004, with the goal of preventing the international spread of contaminated food and foodborne disease and strengthening food safety systems globally. This is done by:

1. Promoting the rapid exchange of information during food safety events;
2. Sharing information on important food safety issues of global interest;
3. Promoting partnership and collaboration between countries; and
4. Helping countries strengthen their capacity to manage food safety risks.

Membership to INFOSAN is voluntary, but is restricted to representatives from national and regional government authorities and requires an official letter of designation. INFOSAN seeks to reflect the multidisciplinary nature of food safety and promote intersectoral collaboration by requesting the designation of Focal Points in each of the respective national authorities with a stake in food safety, and a single Emergency Contact Point in the national authority with the responsibility for coordinating national food safety emergencies; countries choosing to be members of INFOSAN are committed to sharing information between their respective food safety authorities and other INFOSAN members.

The operational definition of a food safety authority includes those authorities involved in: food policy; risk assessment; food control and management; food inspection services; foodborne disease surveillance and response; laboratory services for monitoring and surveillance of foods and foodborne diseases; and food safety information, education and communication across the farm-to-table continuum.

Regulatory Steps

Food may be contaminated during all stages of food production and retailing. In order to prevent viral contamination, regulatory authorities in Europe have enacted several measures:

- European Commission Regulation (EC) No 2073/2005 of November 15, 2005
- European Committee for Standardization (CEN): Standard method for the detection of norovirus and hepatitis A virus in food products (shellfish, fruits and vegetables, surfaces and bottled water)
- CODEX Committee on Food Hygiene (CCFH): Guideline for the application of general principles of food hygiene for the control of viruses in food

Cheese

Cheese is a food derived from milk that is produced in a wide range of flavours, textures, and forms by coagulation of the milk protein casein. It comprises proteins and fat from milk, usually the milk of cows, buffalo, goats, or sheep. During production, the milk is usually acidified, and adding the enzyme rennet causes coagulation. The solids are separated and pressed into final form. Some cheeses have molds on the rind or throughout. Most cheeses melt at cooking temperature.

Hundreds of types of cheese from various countries are produced. Their styles,textures and flavours depend on the origin of the milk (including the animal's diet), whether they have been pasteurized, the butterfat content, the bacteria and mold, the processing, and aging.

Herbs, spices, or wood smoke may be used as flavouring agents. The yellow to red colour of many cheeses, such as Red Leicester, is produced by adding annatto. Other ingredients may be added to some cheeses, such as black peppers, garlic, chives or cranberries.

For a few cheeses, the milk is curdled by adding acids such as vinegar or lemon juice. Most cheeses are acidified to a lesser degree by bacteria, which turn milk sugars into lactic acid, then the addition of rennet completes the curdling.Vegetarian alternatives to rennet are available; most are produced by fermentation of the fungus *Mucor miehei*, but others have been extracted from various species of the *Cynara* thistle family.

Cheese is valued for its portability, long life, and high content of fat, protein, calcium, and phosphorus. Cheese is more compact and has a longer shelf life than milk, although how long a cheese will keep may depend on the type of cheese; labels on packets of cheese often claim that a cheese should be consumed within three to five days of opening. Generally speaking, hard cheeses last longer than soft cheeses, such as Brie or goat's milk cheese.

Cheesemakers near a dairy region may benefit from fresher, lower-priced milk, and lower shipping costs. The long storage life of some cheese, especially if it is encased in a protective rind, allows selling when markets are favourable.

A specialist seller of cheese is sometimes known as a *cheesemonger*. Becoming an expert in this field requires some formal education and years of tasting and hands-on experience, much like becoming an expert in wine or cuisine. The cheesemonger is responsible for all aspects of the cheese inventory: selecting the cheese menu, purchasing, receiving, storage, and ripening.

There is some debate as to the best way to store cheese, but some experts say that wrapping it in cheese paper provides optimal results. Cheese paper is coated in a porous plastic on the inside, and the outside has a layer of wax.

This specific combination of plastic on the inside and wax on the outside protects the cheese by allowing condensation on the cheese to be wicked away while preventing moisture from within the cheese to escape.

Production

Curdling

Figure: *Swiss cheesemaking (heating stage)*

Figure: *During industrial production ofEmmental cheese, the as-yet-undrained curd is broken by rotating mixers.*

A required step in cheesemaking is separating the milk into solid curds and liquidwhey. Usually this is done by acidifying (souring) the milk and adding rennet. The acidification can be accomplished directly by the addition of an acid, such as vinegar, in a few cases (paneer, queso fresco). More commonly starter bacteria are employed instead which convert milk sugars into lactic acid. The same bacteria (and the enzymes they produce) also play a large role in the eventual flavour of aged cheeses. Most cheeses are made with starter bacteria from the *Lactococcus,Lactobacillus*, or *Streptococcus* families. Swiss starter cultures also include*Propionibacter shermani*, which produces carbon dioxide gas bubbles during aging, giving Swiss cheese or Emmental its holes (called "eyes").

Some fresh cheeses are curdled only by acidity, but most cheeses also use rennet. Rennet sets the cheese into a strong and rubbery gel compared to the fragile curds produced by acidic coagulation alone. It also allows curdling at a lower acidity—important because flavour-making bacteria are inhibited in high-acidity environments. In general, softer, smaller, fresher cheeses are curdled with a greater proportion of acid to rennet than harder, larger, longer-aged varieties.

While rennet was traditionally produced via extraction from the inner mucosa of the fourth stomach chamber of slaughtered young, unweaned calves, most rennet used today in cheesemaking is produced recombinantly. The majority of the applied chymosin is retained in the whey and, at most, may be present in cheese in trace quantities. In ripe cheese, the type and provenance of chymosin used in production cannot be determined.

Curd Processing

At this point, the cheese has set into a very moist gel. Some soft cheeses are now essentially complete: they are drained, salted, and packaged. For most of the rest, the curd is cut into small cubes. This allows water to drain from the individual pieces of curd.

Some hard cheeses are then heated to temperatures in the range of 35–55 °C (95–131 °F). This forces more whey from the cut curd. It also changes the taste of the finished cheese, affecting both the bacterial culture and the milk chemistry. Cheeses that are heated to the higher temperatures are usually made with thermophilic starter bacteria that survive this step—either *Lactobacilli* or *Streptococci*.

Salt has roles in cheese besides adding a salty flavour. It preserves cheese from spoiling, draws moisture from the curd, and firms cheese's texture in an interaction with its proteins. Some cheeses are salted from the outside with dry salt or brine washes. Most cheeses have the salt mixed directly into the curds.

Figure: *Cheese factory in the Netherlands*

Other techniques influence a cheese's texture and flavour. Some examples are :

- Stretching: (Mozzarella, Provolone) The curd is stretched and kneaded in hot water, developing a stringy, fibrous body.
- Cheddaring: (Cheddar, other English cheeses) The cut curd is repeatedly piled up, pushing more moisture away. The curd is also mixed (or *milled*) for a long time, taking the sharp edges off the cut curd pieces and influencing the final product's texture.
- Washing: (Edam, Gouda, Colby) The curd is washed in warm water, lowering its acidity and making for a milder-tasting cheese.

Most cheeses achieve their final shape when the curds are pressed into a mold or form. The harder the cheese, the more pressure is applied. The pressure drives out moisture—the molds are designed to allow water to escape—and unifies the curds into a single solid body.

Figure: *Parmigiano-Reggiano in a modern factory*

Ripening

A newborn cheese is usually salty yet bland in flavour and, for harder varieties, rubbery in texture. These qualities are sometimes enjoyed—cheese curds are eaten on their own—but normally cheeses are left to rest under controlled conditions. This aging period (also called ripening, or, from the French, *affinage*) lasts from a few days to several years. As a cheese ages, microbes and enzymes transform texture and intensify flavour. This transformation is largely a result of the breakdown of casein proteins and milkfat into a complex mix of amino acids, amines, and fatty acids.

Some cheeses have additional bacteria or molds intentionally introduced before or during aging. In traditional cheesemaking, these microbes might be already present in the aging room; they are simply allowed to settle and grow on the stored cheeses. More often today, prepared cultures are used, giving more consistent results and putting fewer constraints on the environment where the cheese ages. These cheeses include soft ripened cheeses such as Brie andCamembert, blue cheeses such as Roquefort, Stilton, Gorgonzola, and rind-washed cheeses such as Limburger.

Types

Figure: *Feta from Greece*

There are many types of cheese, with around 500 different varieties recognized by the International Dairy Federation, more than 400 identified by Walter and Hargrove, more than 500 by Burkhalter, and more than 1,000 by Sandine and Elliker. The varieties may be grouped or classified into types according to criteria such as length of ageing, texture, methods of making, fat content, animal milk, country or region of origin, etc.—with these criteria either being used singly or in combination, but with no single method being universally used. The method most commonly and traditionally used is based on moisture content, which is then further discriminated by fat content and curing or ripening methods. Some attempts have been made to rationalise the classification of cheese—a scheme was proposed by Pieter Walstra which uses the primary and secondary starter combined with moisture content, and Walter and Hargrove suggested classifying by production methods which produces 18 types, which are then further grouped by moisture content.

Moisture Content (Soft to Hard)

Categorizing cheeses by firmness is a common but inexact practice. The lines between "soft", "semi-soft", "semi-hard", and "hard" are arbitrary, and many types of cheese are made in softer or firmer variations. The main factor that controls cheese hardness is moisture content, which depends largely on the pressure with which it is packed into molds, and on aging time.

Fresh, Whey and Stretched Curd Cheeses

The main factor in the categorization of these cheeses is their age. Fresh cheeses without additional preservatives can spoil in a matter of days.

Content (Double Cream, Goat, Ewe and Water Buffalo)

Figure: *Emmental*

Some cheeses are categorized by the source of the milk used to produce them or by the added fat content of the milk from which they are produced. While most of the world's commercially available cheese is made from cows' milk, many parts of the world also produce cheese from goats and sheep. Double cream cheeses are soft cheeses of cows' milk enriched with cream so that their fat content is 60% or, in the case of triple creams, 75%.

Soft-ripened and Blue-vein

There are at least three main categories of cheese in which the presence of mold is a significant feature: soft ripened cheeses, washed rind cheeses and blue cheeses.

Processed cheeses

Processed cheese is made from traditional cheese and emulsifying salts, often with the addition of milk, more salt,preservatives, and food colouring. It is inexpensive, consistent, and melts smoothly. It is sold packaged and either pre-sliced or unsliced, in a number of varieties. It is also available in aerosol cans in some countries.

Eating and Cooking

At refrigerator temperatures, the fat in a piece of cheese is as hard as unsoftenedbutter, and its protein structure is stiff as well. Flavour and odour compounds are less easily liberated when cold. For improvements in flavour and texture, it is widely advised that cheeses be allowed to warm up to room temperature before eating. If the cheese is further warmed, to 26–32 °C (79–90 °F), the fats will begin to "sweat out" as they go beyond soft to fully liquid.

Figure: *Zigerbrüt, cheese grated onto bread through a mill, from the Canton of Glarus in Switzerland.*

Above room temperatures, most hard cheeses melt. Rennet-curdled cheeses have a gel-like protein matrix that is broken down by heat. When enough protein bonds are broken, the cheese itself turns from a solid to a viscous liquid. Soft, high-moisture cheeses will melt at around 55 °C (131 °F), while hard, low-moisture cheeses such as Parmesan remain solid until they reach about 82 °C (180 °F).Acid-set cheeses, including halloumi, paneer, some whey cheeses and many varieties of fresh goat cheese, have a protein structure that remains intact at high temperatures. When cooked, these cheeses just get firmer as water evaporates.

Some cheeses, like raclette, melt smoothly; many tend to become stringy or suffer from a separation of their fats. Many of these can be coaxed into melting smoothly in the presence of acids or starch. Fondue, with wine providing the acidity, is a good example of a smoothly melted cheese dish. Elastic stringiness is a quality that is sometimes enjoyed, in dishes includingpizza and Welsh rarebit. Even a melted cheese eventually turns solid again, after enough moisture is cooked off. The saying "you can't melt cheese twice" (meaning "some things can only be done once") refers to the fact that oils leach out during the first melting and are gone, leaving the non-meltable solids behind.

As its temperature continues to rise, cheese will brown and eventually burn. Browned, partially burned cheese has a particular distinct flavour of its own and is frequently used in cooking (e.g., sprinkling atop items before baking them).

Health and Nutrition

The nutritional value of cheese varies widely. Cottage cheese may consist of 4% fat and 11% protein; some whey cheeses 15% fat and 11% protein, and some triple-crème cheeses 36% fat and 7% protein. In general, cheese supplies a great deal of calcium, protein, phosphorus and fat. A 30-gram (1.1 oz) serving of Cheddar cheese contains about 7 grams (0.25 oz) of protein and 200 milligrams of calcium. Nutritionally, cheese is essentially concentrated milk: it takes about 200 grams (7.1 oz) of milk to provide that much protein, and 150 grams (5.3 oz) to equal the calcium.

Pasteurization

A number of food safety agencies around the world have warned of the risks of raw-milk cheeses. The U.S. Food and Drug Administration states that soft raw-milk cheeses can cause "serious infectious diseases including listeriosis, brucellosis, salmonellosis and tuberculosis". It is U.S. law since 1944 that all raw-milk cheeses (including imports since 1951) must be aged at least 60 days. Australia has a wide ban on raw-milk cheeses as well, though in recent years exceptions have been made for Swiss Gruyère, Emmental and Sbrinz, and for French Roquefort.There is a trend for cheeses to be pasteurized even when not required by law.

Compulsory pasteurization is controversial. Pasteurization does change the flavour of cheeses, and unpasteurized cheeses are often considered to have better flavour, so there are reasons not to pasteurize all cheeses. Some say that health concerns are overstated, or that milk pasteurization does not ensure cheese safety.

Pregnant women may face an additional risk from cheese; the U.S. Centres for Disease Control has warned pregnant women against eating soft-ripened cheeses and blue-veined cheeses, due to the listeria risk, which can cause miscarriage or harm to the fetus during birth.

World Production and Consumption

Worldwide, cheese is a major agricultural product. According to the Food and Agricultural Organization of the United Nations, over 20 million metric tons of cheese were produced worldwide in 2011. This is about three kilograms for each person on Earth.

The largest producer of cheese is the United States, accounting for 26% of world production, followed by Germany and France. In the U.S., mozzarella and cheddar are by far the top cheese types.

Yogurt

Yogurt, yoghurt, or yoghourt is a food produced by bacterial fermentation ofmilk. The bacteria used to make yogurt are known as "yogurt cultures". Fermentation of lactose by these bacteria produces lactic acid, which acts on milk protein to give yogurt its texture and its characteristic tang.

Worldwide, cow's milk, the protein of which is mainly casein, is most commonly used to make yogurt. Milk from water buffalo, goats, ewes, mares, camels, and yaks however, is also used to produce yogurt in various parts of the world.

Dairy yogurt is produced using a culture of Lactobacillus delbrueckii subsp.bulgaricus and Streptococcus thermophilus bacteria. In addition, otherlactobacilli and bifidobacteria are also sometimes added during or after culturing yogurt. Some countries require yogurt to contain a certain amount of colony-forming units of microorganisms.

In Western culture, the milk is first heated to about 85 °C (185 °F) to denaturethe milk proteins so that they set together rather than form curds. In some places, such as parts of India and Bangladesh, curds are a desired component and milk is boiled. After heating, the milk is allowed to cool to about 45 °C (113 °F). The bacterial culture is added, and the temperature of 45 °C is maintained for 4 to 7 hours to allow fermentation.

Nutritional Value

Yogurt is nutritionally rich in protein, calcium, vitamin D, riboflavin, vitamin B6, andvitamin B12. It has nutritional benefits beyond those of milk. Lactose-intolerantindividuals may tolerate yogurt better than other dairy products due to the conversion of lactose to the sugars glucose and galactose, and due to the fermentation of lactose to lactic acid carried out by the bacteria present in the yogurt. Yogurt contains varying amounts of fat. For example, some cows'-milk yogurts contain no fat; others of low fat content have 2% fat, whole-milk yogurt may have 4% fat; some yogurts sold as "Greek-style" may have about 10% fat.

Figure: *Unstirred Turkish* Süzme Yoðurt*(strained yogurt), with a 10% fat content*

Health Effects

Yogurt has been claimed to have many health benefits. There is moderate-quality evidence to support the idea that consumption of dairy products, including yogurt, may reduce the risk of high blood pressure. However, the precise mechanism for this effect is not fully understood.

Varieties and Presentation

Da-hi is a yogurt of the Indian subcontinent, known for its characteristic taste and consistency. The word *da-hi* seems to be derived from the Sanskritword *dadhi*, one of the five elixirs, or panchamrita, often used in Hindu ritual. *Dahi* also holds cultural symbolism in many homes in the *Mithila* region ofNepal and Bihar. Yogurt balances the palate across regional cuisines throughout India. In the hot and humid south, yogurt and foods made of yogurt are a staple in order to cool down - yogurt rice is always the last dish of the meal. Also, the primarily vegetarian population of India derives some protein from yogurt (other than lentil and beans). Sweet yogurt (*meesti doi* or *meethi dahi*) is common in eastern parts of India, made by fermenting sweetened milk. While cows' milk is considered sacred and is currently the primary ingredient for yogurt, goat and buffalo milk were widely used in the past, and valued for the fat content. Butter and cream were made by churning the yogurt/milk.

In India and Pakistan, it is often used in cosmetics mixed with turmeric and honey. Sour yogurt, is also used as a hair conditioner by women in many parts of India and Pakistan. *Dahi* is also known as *Mosaru* (Kannada), *Thayir* (Tamil), *Thayiru* (Malayalam), *doi* (Assamese, Bengali), *dohi* (Oriya), *perugu* (Telugu), and *Dhahi* or *Dhaunro* (Sindhi).

Raita is a yogurt-based South Asian/Indian condiment, used as a side dish. The yogurt is seasoned with cilantro (coriander), cumin, mint, cayenne pepper, and other herbs and spices. Vegetables such as cucumber and onions are mixed in, and the mixture is served chilled. Raita has a cooling effect on the palate which makes it a good foil for spicy Indian and Pakistani dishes. Raita is sometimes also referred to as *dahi*.

Dadiah or Dadih is a traditional West Sumatran yogurt made from water buffalo milk, fermented in bamboo tubes. Yogurt is popular in Nepal, where it is served as both an appetizer and dessert. Locally called *dahi* , it is a part of the Nepali culture, used in local festivals, marriage ceremonies, parties, religious occasions, family gatherings, and so on. The most famous type of Nepalese yogurt is called *juju dhau*, originating from the city of Bhaktapur. In Tibet, yak milk (technically dri milk, as the word yak refers to the male animal) is made into yogurt (and butter and cheese) and consumed.

In Northern Iran, *Mßst Chekide* is a variety of kefir yogurt with a distinct sour taste. It is usually mixed with a pesto-like water and fresh herb purée called delal. Yogurt is a side dish to all Iranian meals. The most popular appetizers are spinach oreggplant borani, *Mßst-o-Khiâr* with cucumber, spring onions and herbs, and *Mâst-Musir* with wild shallots. In the summertime, yogurt and ice cubes are mixed together with cucumbers, raisins, salt, pepper and onions and topped with some croutons made of Persian traditional bread and served as a cold soup. Ashe-Mâst is a warm yogurt soup with fresh herbs, spinach and lentils. Even the leftover water extracted when straining yogurt is cooked to make a sour cream sauce called kashk, which is usually used as a topping on soups and stews.

Matsoni is a Georgian yogurt popular in the Caucasus and Russia. It is used in a wide variety of Georgian dishes and is believed to have contributed to the high life expectancy and longevity in the country. Dannon used this theory in their 1978 TV advertisement called *In Soviet Georgia* where shots of elderly Georgian farmers were interspersed with an off-cameraannouncer intoning, "In Soviet Georgia, where they eat a lot of yogurt, a lot of people live past 100." Matsoni is also popular in Japan under the name Caspian Sea Yogurt.

Tarator and Cacýk are popular cold soups made from yogurt, popular during summertime in Albania, Azerbaijan (known as Dogramac), Bulgaria, Macedonia, Serbia and Turkey. They are made with ayran, cucumbers, dill, salt, olive oil, and optionally garlic and ground walnuts. Tzatziki in Greece and milk salad in Bulgaria are thick yogurt-based salads similar to tarator.

Khyar w Laban (cucumber and yogurt salad) is a popular dish in Lebanon and Syria. Also, a wide variety of local Lebanese and Syrian dishes are cooked with yogurt like "Kibbi bi Laban", etc.

Rahmjoghurt, a creamy yogurt with much higher fat content (10%) than many yogurts offered in English-speaking countries (*Rahm* is German for "cream"), is available in Germany and other countries.

Dovga, a yogurt soup cooked with a variety of herbs and rice is popular in Azerbaijan, often served warm in winter or refreshingly cold in summer.

Cream-top yogurt is yogurt made with unhomogenized milk. A layer of cream rises to the top, forming a rich yogurt cream. Cream-top yogurt was first made commercially popular in the United States by Brown Cow of Newfield, New York, bucking the trend towards low- and non-fat yogurts.

Jameed is yogurt which is salted and dried to preserve it. It is popular in Jordan.

Zabadi is the type of yogurt made in Egypt, usually from the milk of the Egyptian water buffalo. It is particularly associated with Ramadan fasting, as it is thought to prevent thirst during all-day fasting.

Sweetened and Flavoured Yogurt

To offset its natural sourness, yogurt is also sold sweetened, flavoured or in containers with fruit or fruit jam on the bottom. The two styles of yogurt commonly found in the grocery store are set type yogurt and Swiss style yogurt. Set type yogurt is when the yogurt is packaged with the fruit on the bottom of the cup and the yogurt on top. Swiss style yogurt is when the fruit is blended into the yogurt prior to packaging.

Lassi and Moru are common beverages in India. Lassi is milk that is sweetened with sugar commonly, less commonly honey and often combined with fruit pulp to create flavoured lassi. Mango lassi is a western favourite, as is coconut lassi. Consistency can vary widely, with urban and commercial lassis being of uniform texture through being processed, whereas rural and rustic lassi has curds in it, and sometimes has malai (cream) added or removed. Moru is a popular South Indian summer drink, meant to keep drinkers hydrated through the hot and humid summers of the South. It is prepared by considerably thinning down yogurt with water, adding salt (for electrolyte balance) and spices, usually green chilli peppers, asafoetida, curry leaves and mustard.

Large amounts of sugar – or other sweeteners for low-calorie yogurts – are often used in commercial yogurt. Some yogurts contain added starch, pectin (found naturally in fruit), and/or gelatin to create thickness and creaminess artificially at lower cost. This type of yogurt is also marketed under the name Swiss-style, although it is unrelated to the way yogurt is eaten in Switzerland. Some yogurts, often called "cream line," are made with whole milk which has not been homogenized so the cream rises to the top. Fruit jam is used instead of raw fruit pieces in fruit yogurts to allow storage for weeks.

In the UK, Ireland, France and USA, sweetened, flavoured yogurt is the most popular type, typically sold in single-serving plastic cups. Common flavours include vanilla, honey, and toffee, and fruit such as strawberry, cherry, blueberry, blackberry,raspberry, mango and peach. In the early twenty-first century yogurt flavours inspired by desserts, such as chocolate or cheesecake, have been available.

Strained Yogurt

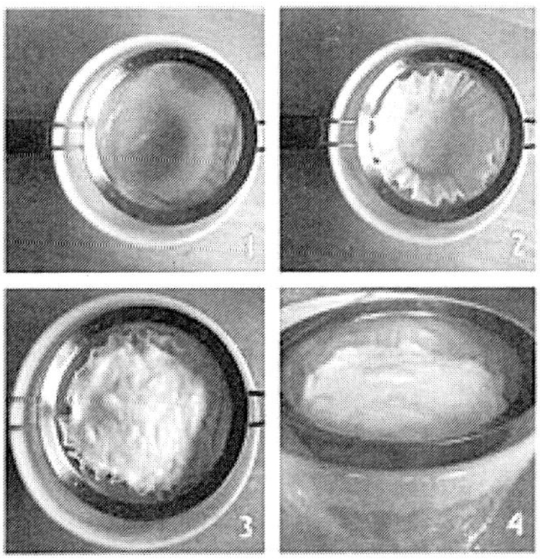

Figure: *Use coffee filter to strain yogurt in a home refrigerator.*

Strained yogurt is yogurt which has been strained through a paper or cloth filter, traditionally made of muslin., to remove the whey, giving a much thicker consistency and a distinctive, slightly tangy taste. Strained yogurt is becoming more popular with those who make yogurt at home, especially if using skim milk which results in a thinner consistency. Once yogurt is made and refrigerated overnight, it is poured in a muslin or cheesecloth bag and hung in the coolest place in the house, with a tub placed underneath to collect the dripping whey. In cold weather a single day (or night) of straining is sufficient. In higher ambient temperatures yogurt will spoil rapidly, therefore it had best be actively squeezed or strained until about a third or more of its initial weight has run off. The remainder is now strained and is refrigerated again.

Labneh is a strained yogurt used for sandwiches popular in Arab countries. Olive oil, cucumber slices, olives, and various green herbs may be added. It can be thickened further and rolled into balls, preserved in olive oil, and fermented for a few more weeks. It is sometimes used with onions, meat, and nuts as a stuffing for a variety of pies or kebbeh (ßÈÉ) balls.

Some types of strained yogurts are boiled in open vats first, so that the liquid content is reduced. The popular East Indian dessert, a variation of traditional dahi called mishti dahi, offers a thicker, more custard-like consistency, and is usually sweeter than western yogurts.

Strained yogurt is also enjoyed in Greece and is the main component of *tzatziki* (from Turkish "cacýk"), a well-known accompaniment to gyros and souvlaki pita sandwiches: it is a yogurt sauce or dip made with the addition of gratedcucumber, olive oil, salt and, optionally, mashed garlic.

Srikhand, a popular dessert in India, is made from strained yogurt, saffron, cardamom, nutmeg and sugar and sometimes fruits such as mango or pineapple.

In North America and Britain, strained yogurt is commonly called "Greek yogurt".

Beverages

Dugh ("dawghe" in Neo-Aramaic), ayran or dhallë is a yogurt-based, salty drink popular in Iran, Kurdistan Albania, Bulgaria, Turkey, Azerbaijan, Afghanistan, Pakistan, Bangladesh, Macedonia, Kazakhstan and Kyrgyzstan. It is made by mixing yogurt with water and (sometimes) salt. The same drink is known as *doogh* in Iran and in some parts of Kurdistan; *tan* in Armenia; *laban ayran* in Syria and Lebanon; *shenina* in Iraq and Jordan; *laban arbil* in Iraq; *majjiga*

(Telugu),*majjige* (Kannada), and *moru* (Tamil andMalayalam) in South India; namkeen *lassi* inPunjab and all over Pakistan. A similar drink,doogh, is popular in the Middle East between Lebanon, Iran, and Iraq; it differs from ayran by the addition of herbs, usuallymint, and is sometimes carbonated, commonly with carbonated water.

Figure: *PCC Dairy Yogurt Milk, with live cultures, made from water buffalo's cream milk Philippine Carabao Centre.*

Figure: *Ayran. One of the popular beverages in Turkish cuisine.*

Borhani (or Burhani) is a spicy yogurt drink popular in Bangladesh and parts of Bengal. It is usually served with kacchibiryani at weddings and special feasts. Key ingredients are yogurt blended with mint leaves (mentha), mustard seeds and black rock salt (Kala Namak). Ground roasted cumin, ground white pepper, green chili pepper paste and sugar are often added.

Lassi is a yogurt-based beverage originally from the Indian subcontinent that is usually slightly salty or sweet. Lassi is a staple in Punjab. In some parts of the subcontinent, the sweet version may be commercially flavoured with rosewater, mango or other fruit juice to create a very different drink. Salty lassi is usually flavoured with ground, roastedcumin and red chilies; this salty variation may also use buttermilk, and in India is interchangeably called *ghol* (Bengal), *mattha* (North India), "majjige" (Karnataka), *majjiga* (Telangana & Andhra Pradesh), *moru* (Tamil Nadu and Kerala), *Dahi paani Chalha* (Odisha), *tak* (Maharashtra), or *chaas* (Gujarat). Lassi is very widely drunk in India, Pakistan, andBangladesh. Mango Lassi is a popular drink at Indian restaurants in US.

In Bosnia and Herzegovina, Croatia, Macedonia, Montenegro, Serbia, and Slovenia, an unsweetened and unsalted yogurt drink usually called simply *jogurt* is a popular accompaniment to *burek* and other baked goods. Sweetened yogurt drinks are the usual form in Europe (including the UK) and the US, containing fruit and added sweeteners. These are typically called "drinking / drinkable yogurt", such as Yop and BioBest Smoothie. Also available are "yogurt smoothies", which contain a higher proportion of fruit and are more like smoothies. In Ecuador, yogurt smoothies flavoured with native fruit are served with pan de yuca as a common type of fast food. Also in Turkey, yogurt-soup or *Yayla Çorbasý* is a popular way of consuming yogurt. The soup is a mix of yogurt, rice, flour and dried mint.

Making Yogurt at Home

Yogurt is made by inoculating certain bacteria (starter culture), usually *Streptococcus thermophilus* and *Lactobacillus bulgaricus*, into milk. Starter culture may be provided by a pre-packaged powder or an amount of existing yogurt with live bacteria. After inoculation, the milk is incubated at 40 to 46 °C (105 to 115 °F) until firm; the milk is coagulated by bacteria-produced lactic acid.

The milk used to make yogurt contains a higher concentration of solids than normal milk. By increasing the solids content of the milk, a firm, rather than soft, end product results. Addition of nonfat dry milk (NFDM) is the easiest at-home method for doing this. Another method is using scalded milk: heating milk near boiling, then letting it cool down to 40 to 46 °C (105 to 115 °F). The process denaturates whey proteins, which makes yogurt thicker.

The yogurt making process provides two significant barriers to pathogen growth: (a) heat and (b) acidity (low pH). Both are necessary to ensure a safe product. Acidity alone has been questioned by recent outbreaks of food poisoning by *E. coli O157:H7* that is acid-tolerant. *E. coli O157:H7* is easily destroyed by pasteurization (heating). Therefore, pasteurized milk is used for making yogurt.

This initial process is the first step in making strained yogurt as mentioned above by eliminating some of the liquid whey.

Non-dairy Yogurt Substitutes

Various manufacturers have endeavoured to produce plant-based substitutes for yogurt, using soy milk, rice milk, nut milkssuch as almond milk, and coconut milk, for consumers who are unable to tolerate dairy products in any form.Although Lactobacillus strains exist that can produce lactic acid from plant-derived feed stock (e.g. sauerkraut), since plants do not produce lactose, it is not possible to produce genuine yogurt from plant sources with the same bacteria cultures used to convert milk into yogurt.

Frying Fats

Frying is the cooking of food in oil or another fat, a technique that originated in ancient Egypt around 2500 BC.Chemically, oils and fats are the same, differing only in melting point, and the distinction is only made when needed. Foods can be fried in a variety of fats, including lard, vegetable oil, rapeseed oil and olive oil. In commerce, many fats are called oils by custom, e.g. palm oil and coconut oil, which are solid at room temperature. A variety of foods may be fried, including the potato chip, bread, eggs and foods made from eggs, such as omelettes or pancakes.

Details

Fats can reach much higher temperatures than water at normal atmospheric pressure. Through frying, one can sear or even carbonize the surface of foods whilecaramelizing sugars. The food is cooked much more quickly and has a characteristic crispness and texture. Depending on the food, the fat will penetrate it to varying degrees, contributing richness, lubricity, and its own flavour, as well as calories.

Frying techniques vary in the amount of fat required, the cooking time, the type of cooking vessel required, and the manipulation of the food. Sautéing, stir frying, pan frying, shallow frying, and deep frying are all standard frying techniques.

Figure: *Polenta, French fries, and fried chicken at a Brazilian eatery*

Pan frying, sautéing and stir-frying involve cooking foods in a thin layer of fat on a hot surface, such as a frying pan, griddle, wok, or sauteuse. Stir frying involves frying quickly at very high temperatures, requiring that the food be stirred continuously to prevent it from adhering to the cooking surface and burning.

Shallow frying is a type of pan frying using only enough fat to immerse approximately one-third to one-half of each piece of food; fat used in this technique is typically only used once. Deep-frying, on the other hand, involves totally immersing the food in hot oil, which is normally topped up and used several times before being disposed. Deep-frying is typically a much more involved process, and may require specialized oils for optimal results.

Deep frying is now the basis of a very large and expanding worldwide industry. Fried products have consumer appeal in all age groups and in virtually all cultures, and the process is quick, can easily be made continuous for mass production, and the food emerges sterile and dry, with a relatively long shelf life. The end products can then be easily packaged for storage and distribution. Examples are potato chips, french fries, nuts, doughnuts, instant noodles, etc.

Shallow (Pan) Frying

Shallow frying is an oil-based cooking technique. It is typically used to prepare portion-sized cuts of meat and fish, and patties such as fritters. Shallow frying can also be used to cook vegetables. Shallow-fried foods are often battered. It is a high-heat process, promoting browning and, in some cases, a Maillard reaction. Since the food is only partly submerged, it must be turned over partway through the cooking process. Some cooks recommend cooking the "presentation" side of the food first, since this side will be better browned.

Deep Fat Frying

Deep frying (also referred to as deep fat frying) is a cooking method in which food is submerged in hot fat (e.g., oil). This is normally performed with a deep fryer or chip pan; industrially, a pressure fryer or vacuum fryermay be used.

Deep frying is classified as a dry cooking method because no water is used. Due to the high temperature involved and the high heat conduction of oil, it cooks food extremely quickly.

***Figure:** Czech bramboraèek bread being deep fried*

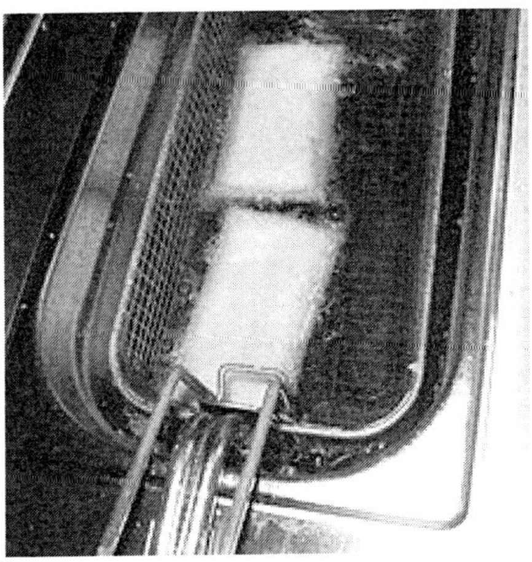

***Figure:** Deep frying kaassoufflés in a deep fryer*

Technique

If performed properly, deep-frying does not make food excessively greasy, because the moisture in the food repels the oil. The hot oil heats the water within the food,steaming it; oil cannot go against the direction of this powerful flow because (due to its high temperature) the water vapor pushes the bubbles towards the surface. As long as the oil is hot enough and the food is not immersed in the oil for too long, oil penetration will be confined to the outer surface. However, if the food is cooked in the oil for too long, much of the water will be lost and the oil will begin to penetrate the food. The correct frying temperature depends on the thickness and type of food, but in most cases it lies between 175–190 °C (347–374 °F).

Oil Deterioration

Overheating or over-using the frying oil leads to formation of rancid-tasting products of oxidation, polymerization, and other deleterious, unintended or even toxic compounds such as acrylamide (from starchy foods). Deep-frying under vacuumhelps to significantly reduce acrylamide formation, but this process is not widely used in the food industry due to the high investment cost involved.

Some useful tests and indicators of excessive oil deterioration are the following:

- Sensory: Darkening, smoke, foaming, thickening, rancid taste and unpleasant smell when heating.
- Laboratory: Acidity, anisidine value, viscosity, total polar compounds, polymeric triglycerides.

Instruments that indicate total polar compounds, currently the best single gauge of how deep-fried an object is, are available with sufficient accuracy for restaurant and industry use.

Bibliography

Ashraf, Syed Mahboob: *Handbook of Fruit and Vegetable Products*, Agrobios, Delhi, 2008.

Bajpai, Amita : *Recent Trends in Animal Nutrition and Feed Technology for Livestock, Pets and Laboratory Animals*, International, 2003.

Bandana Ghosh, M.S. Ranganathan and Sriram Sridhar : *Enzyme and Food Biotechnology*, Wisdom Press, 2011.

Brij K. Taimni: *Food Security in 21 Century : Perspective and Vision*, Konark, Delhi, 2001.

Chakraborty, Sudip : *Food Security and Child Labour : The Case of a Hazardous Occupation*, Deep and Deep, Delhi, 2011.

Cristobal Noe Aguilar : *Food Science and Food Biotechnology in Developing Countries*, Asiatech Pub, 2008.

Daphne C. Elliott: *Biochemistry and Molecular Biology*, Oxford University Press, Delhi, 2005.

Dhaka, J.M. : *Economics of Agricultural Production and Farm Management*, Aavishkar Pub, Delhi, 2009.

Fox, Patrick F.: *Advanced Dairy Chemistry: Proteins*, New York: Elsevier Applied Science, 1992.

Gutfreund, H. : *Enzymes: Physical Principles*, Wiley-Interscience, London, 1972.

Harsh Bhaskar: *Problems in Biochemistry*, Campus Books, Delhi, 2010.

Jack Simons: *An Introduction to Theoretical Chemistry*, Cambridge Univ Press, Delhi, 1998.

Jagbir Sharma: *Advanced Environment Chemistry*, RBSA Pub, Delhi, 2006.

James L. Steele: *Applied Dairy Microbiology*, New York: M. Dekker, 1998.

John Prince: *Dairy Farming: Being the Theory, Practice, and Methods of Dairying*, New York, 1888.

Lata Bhattacharya: *Animal Biochemistry*, Discovery, Delhi, 2010.

Mayrose, V.B.: *Handbook of livestock Management Techniques*, Macmillan Publishing Co., New York, 1981.

Mowery, D. C. : *Technology and Wealth of Nations*, Stanford: Stanford University Press, 1992.

Narang, R.K.: *Fruit and Vegetable Preservation Techniques*, APH Pub, Delhi, 2010.

Paul B. : *Food Biotechnology in Ethical Perspective*, Aspen, CO: Aspen Publishers, 1997.

Qystein V. Sjaastad: *Physiology of Domestic Animals*, International Book Distributing Co., Delhi, 2005.

Raghunath Narvekar: *Handbook of Biochemistry*, Adhyayan, Delhi, 2008.

Rao, M K: *Food and Dairy Microbiology*, Manglam Pub, Delhi, 2007.

Schulz, A. R. : *Enzyme Kinetics*, Cambridge University Press, Cambridge, 1994.

Sheldon, John Prince: *Dairy Farming: Being the Theory, Practice, and Methods of Dairying*, New York, 1888.

Sommers, C.H. and Fan, X. : *Food Irradiation Research and Technology*, Blackwell Publishing, Ames, IA, 2006.

Spreer, Edgar : *Milk and Dairy Product Technology*, Translated by Axel Mixa. New York: M. Dekker, 1998.

Sujata K. Dass: *Biotechnology and Food Security*, Isha Books, Delhi, 2004.

Tyagi, Prasum: *A Textbook of Animal Physiology*, Dominant, Delhi, 2010.

Valli Manickam: *A Textbook of Analytical Chemistry*, PharmaMed Press, Delhi, 2006.

Vikram V. Mistry: *Cheese and FermentedMilk Foods*, Great Falls, Va.: Kosikowski, 1997.

Walsh, C. T. : *Enzymatic Reaction Mechanisms*, W.H. Freeman, New York, 1979.

Werbach M.R. : *Nutrition Influences on Mental Illness*, Third Line Press, 1999.

Yadav, D.S.: *Advances in Vegetable Production*, Researchco Book Centre, Delhi, 2005.

Index